Evolution of the Ammonoids

Ammonites are an extinct and charismatic lineage that persisted for over 300 million years. They were used, with other fossils, to corroborate the principle of faunal succession and launch the field of biostratigraphy. Despite intense research, many important questions remain unanswered. Furthermore, outdated hypotheses persist. Many new findings include a better understanding of their appearance in life, their locomotion, and their role in long-gone ecosystems. Of course, there are still controversies, e.g., why did shell complexity increase during evolutionary history. This richly illustrated book describes the full range of ammonoids and their fascinating evolutionary history.

Key Features:

- Documents the early history of paleontology and the role played by ammonoids
- Describes the basic anatomy of a diverse and long-persisting lineage
- Summarizes the classification and diversity of ammonoids
- Lavishly illustrated with beautiful reconstructions
- Highlights recent findings and outstanding controversies

INTERNATIONAL CHRONOSTRATIGRAPHIC CHART

IUGS

www.stratigraphy.org

International Commission on Stratigraphy

v 2023/04

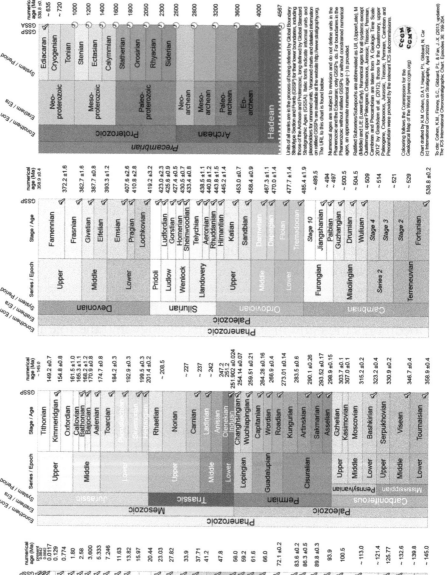

Units of all ranks are in the process of being defined by Global Boundary Stratotype Section and Points (GSSP's) for their lower boundaries, including those of the Archean and Proterozoic, long defined by Global Standard Stratigraphic Ages (GSSA). Italic fonts indicate informal units and placeholders for un-named units. Versioned charts and detailed information on ratified GSSP's are available at the website http://www.stratigraphy.org The URL to this chart is found below.

Numerical ages are subject to revision and do not define units in the Phanerozoic and the Ediacaran; only GSSP's do. For boundaries in the Phanerozoic without ratified GSSP's or without constrained numerical ages, an approximate numerical age (~) is provided.

Ratified Subseries/Subepochs are abbreviated as U/L (Upper/Late), M (Middle) and L/E (Lower/Early). Numerical ages for all systems except Quaternary, Upper Paleogene, Cretaceous, Ordovician and Cambrian are taken from 'A Geologic Time Scale 2012' by Gradstein et al. (2012); those for the Quaternary, upper Paleogene, Cretaceous, Jurassic, Triassic, Permian, Cambrian and Precambrian were provided by the relevant ICS subcommissions.

Colouring follows the Commission for the Geological Map of the World (www.ccgm.org)

Chart drafted by K.M. Cohen, D.A.T. Harper, P.L. Gibbard, N. Car
The ICS International Chronostratigraphic Chart. Episodes 36: 199-204

To cite: Cohen, K.M., Finney, S.C., Gibbard, P.L & Fan, J.-X. (2013; updated)

URL: http://www.stratigraphy.org/ICSchart/ChronostratChart2023-04.pdf

(c) International Commission on Stratigraphy, April 2020

Evolution of the Ammonoids

Kate LoMedico Marriott
With contributions by
Donald R. Prothero and Alexander Bartholomew

CRC Press
Taylor & Francis Group
Boca Raton London New York

CRC Press is an imprint of the
Taylor & Francis Group, an **informa** business

First edition published 2024
by CRC Press
6000 Broken Sound Parkway NW, Suite 300, Boca Raton, FL 33487-2742

and by CRC Press
4 Park Square, Milton Park, Abingdon, Oxon, OX14 4RN

CRC Press is an imprint of Taylor & Francis Group, LLC

© 2024 Kate LoMedico Marriott, Donald R. Prothero, and Alexander Bartholomew

Library of Congress Cataloging-in-Publication Data
Names: Marriott, Kate LoMedico, author. | Prothero, Donald R., author. |
Bartholomew, Alexander J., author.
Title: Evolution of the ammonoids / Kate LoMedico Marriott ; with
contributions by Donald R. Prothero and Alexander Bartholomew.
Description: First edition. | Boca Raton, FL : CRC Press, 2023. |
Includes bibliographical references and index.
Identifiers: LCCN 2022032424 (print) | LCCN 2022032425 (ebook) |
ISBN 9781032264387 (hbk) | ISBN 9781032264363 (pbk) | ISBN 9781003288299 (ebk)
Subjects: LCSH: Ammonoidea.
Classification: LCC QE807.A5 M275 2023 (print) | LCC QE807.A5 (ebook) |
DDC 564/.53—dc23/eng20221205
LC record available at https://lccn.loc.gov/2022032424
LC ebook record available at https://lccn.loc.gov/2022032425

ISBN: 9781032264387 (hbk)
ISBN: 9781032264363 (pbk)
ISBN: 9781003288299 (ebk)

DOI: 10.1201/9781003288299

Typeset in Garamond
by codeMantra

For Rachel

Contents

Preface

As someone who studies or, more accurately, revels in invertebrates, one of my greatest passions is translating the lives of the creatures that seem the least like us into something we can understand. It is my deep belief that understanding fostered between humans and very non-human animals is beneficial to everything alive on the planet. When it comes to fossils, this translation feels a bit more urgent. So little is currently understood about ammonites, and they are frequently overlooked or worse, homogenized, in natural history exhibits that fail to communicate the individuality of invertebrate species.

Ammonites are separated from human beings by about 65 million years. Despite the fact that we never had the chance to adversely impact them while they lived, they should serve as a cautionary tale to us. Nearly every time ammonoids went extinct (almost or completely), the cause was climate change. The closest living relatives to ammonoids are experiencing yet another climate disaster today; this time, we are the ones driving it, and we can be the ones to stop it if we so choose.

Just as many neocoleoids were born out of the extinction which wiped out the ammonites, this book was born out of a disaster in my own life. Shortly before I began working on it, one of my lifelong best friends passed away. My friend Rachel was passionate about marine conservation. She was incredibly knowledgeable on stingrays and had become involved, at the time of her passing, with an organization that protected mantas. It was in my seventh-grade biology class with Rachel that I first encountered ammonoids. I saved up my allowance money and bought a nautilus shell (not yet knowing that it was wrong to buy seashells) that we marveled at for months. Cephalopods had claimed me as forever theirs, and Rachel was the only person with me when it happened.

While I know she would have loved this book, the loss of Rachel is what drove me to immortalize the things we did together, and to carry out work that had been important to both of us for so long. My hope is that this book inspires empathy between humans and invertebrates, sea creatures, and the planet and, at this critical moment in environmental history, that we may look for a bit of commonality in living things that could not appear more different than us, as my friend Rachel would have done.

Kate LoMedico Marriott

Acknowledgments

I am deeply indebted to my colleagues at Taylor & Francis who gave our book (which is my first) a chance, especially Dr. Chuck Crumly, Neha Bhatt, Kara Roberts, and Assunta Petrone. This book would never have been possible without my incredible friends and mentors, including my best friends, Rachel and Elle, and my mom Laura LoMedico Marriott. I am thankful to my lifelong friends Dr. Gary and Janet Lovett for encouraging my scientific interests all my life, and my second grade teacher Mrs. Colantuono, for exposing me to my first paleoart experience in 1998 (even though my mom did all the work). My two coauthors are two of my greatest mentors, and I have the utmost gratitude for them. I have also learned far more than I ever dreamed possible about ammonoids through the most brilliant mentors and colleagues on the subject I could ask for, in particular J.A. Chamberlain, Jr., J. Basil, J. Slattery, C. Klug, R. Hoffmann, P.D. Ward, R. Shell, D. Peterman, G.A. Bishop, D.E. Seidemann, A. Derman, and R.O. Johnson. The lens through which each of them has taught me to see has enriched my ability to discover more and more about marine invertebrates.

Authors

Kate LoMedico Marriott is a lecturer in Earth and Environmental Sciences. She trained in sculpture at the State University of New York at Purchase, then earned an interdisciplinary B.A. at State University of New York at New Paltz, and an M.S. in Earth and Environmental Sciences at Brooklyn College in New York in 2021. Her thesis was on ammonites. She is the former in-house paleontologist at Astro Gallery of Gems on Fifth Avenue in New York City and the author or coauthor of several publications on ammonites, including one systematizing the paleoart theory of heteromorph ammonites and a new method for fractal analysis of ammonite sutures. She has also published research in fossil mammals. Kate was a Special Mention Finalist in the International Award on Scientific Illustration in Madrid, Spain, for her work on heteromorphs in 2018. In 2020, Kate became the second paleoartist to be inducted into the Ocean Artists Society. She is the co-founder of The Society of Invertebrate Paleoartists.

Donald R. Prothero is Research Associate in Vertebrate Paleontology at the Natural History Museum of Los Angeles County. He taught college geology and paleontology for over 45 years, at such places as Caltech, Columbia University, Vassar College, Occidental College, Knox College, and Pierce College, and currently at Cal Poly Pomona. He earned his M.A., M.Phil., and Ph.D. degrees in Geological Sciences from Columbia University, and a B.A. in Geology and Biology (highest honors, Phi Beta Kappa) from the University of California, Riverside. He is currently the author, coauthor, editor, or coeditor of over 48 books and over 350 scientific papers, including 8 leading geology textbooks and 18 trade books as well as edited symposium volumes and other technical works.

Alexander Bartholomew is an Associate Professor at the State University of New York at New Paltz in the Department of Geology, having taught Stratigraphy and Paleontology for over 15 years. He earned his M.S. and Ph.D. in Geology from the University of Cincinnati and his B.S. in Geology from Union College. Alex's area of expertise is in Upper Silurian and Lower-Middle Devonian stratigraphy and paleoecology, spanning marine, marginal marine, and terrestrial environments in eastern North America. Alex is a corresponding member of the International Subcommission on Devonian Stratigraphy (SDS), an organizer/field excursion leader for the upcoming SDS meeting and field excursion in New York State in 2022, and co-/lead author on several papers in the forthcoming *Bulletins of American Paleontology* special volume set.

Chapter 1

Ammon's Horns and Serpent Stones

Jurassic Park of the Oceans

The science fiction visionary Isaac Asimov once said of the mass appeal given to only a limited sect of paleontology, "Of all the extinct life-forms, dinosaurs are the most popular. Why that should be is not clear." It is true that in the 4.5 billion years that complex life has existed on Earth, humans shared the planet with just a handful of fellow winners in the extinction lottery. In the almost 4.5 billion years leading up to the first humans, members of the genus *Homo*, countless scarcely imaginable beasts roamed every corner of the world, occupying every conceivable niche, and some bore forms completely unlike any creature humans have ever seen. Of these, very few were dinosaurs. Most of them weren't even reptiles. More often than not, they didn't even have a spine.

Cephalopods are one such group of spineless beasts whose fossils show they have endured for more than half a billion years. Today, the cephalopods are familiar to most of us as the octopus, squid, cuttlefish (collectively the coleoids) and the ancient survivor, the chambered nautilus. Another group of cephalopods are the extinct ammonites. While ammonites are not common in feature films and museum exhibits, they are popular and intriguing as found objects across cultures. Ammonites are one of the most widely recognized invertebrate fossils in the world, but they have rarely been understood for what they actually are. These enigmatic stone spirals are commonly seen in museums, gothic churches, jewelry, and as

DOI: 10.1201/9781003288299-1

Figure 1.1   **Diverse assemblages of ammonites are vital to accurate Mesozoic seascapes.**

the anonymous motifs of a million designer products. They have become increasingly popular in crystal shops, as they are so abundant that individual specimens are often without their own scientific value.

Although dinosaurs are usually the only group of fossils that gets much public attention, ammonites were the most abundant, and often, the dominant group in the oceans during the Age of Dinosaurs. If you wanted a complete reconstruction of the Jurassic ecosystem for *Jurassic Park*, it would not be just an island off the coast of Costa Rica, as in the novel and movie. The ocean surrounding it would be full of ammonites of every size and shape, swimming and floating and creeping along the bottom, all dominating the ecological niches that fish and coleoids would take by the Age of Mammals (Figure 1.1).

Ammonites in World Culture

Ammonite shells have long been considered magical and sacred objects in cultures all over the world since ancient times. They are often featured in

Figure 1.2 **(a) Saligrams. (b) Buffalo stones.**

ceremonial artworks, and were sometimes talismans for successful hunts. Ever since the Mesolithic, or Middle Stone Age, ammonite fossils have been considered magic talismans. They were sometimes transported great distances through trade routes. In India, they became a prevalent fixture in Hindu architecture. Black limestone concretions with ammonites embedded in them were collected (Figure 1.2a); these were known as *saligrams* (or *shaligrams* or *salagramas*), and treasured as sacred because of their resemblance to the disc chakra held by the Hindu god Vishnu. The chakra system is the Hindu symbol of absolute completeness, with the eight spokes representing the eight-fold path of deliverance. The radial "spokes" that followed this same pattern are formed by the ribs of these ammonites. Saligrams are mentioned in Sanskrit texts dating back to over 2,200 years ago. Other Hindu poems identify them as some kind of worm.

In ancient China, ammonites were known as *Jiaoshih* or "horn stones" because they resembled coils of a ram's horn. In the eleventh century, Chinese scholar and statesman Su Sung wrote is his book, *Pen Tshao Thu Ching*:

> The stone-serpent appears in rocks besides the rivers flowing into the southern seas. Its shape is like a coiled snake no head or tail-tip. Inside it is empty. Its color is reddish purple. The best ones are those which coil to the left. It also looks like a spiral shell of a conch. We do not know what animal it was which was thus changed into stone.

In New Guinea, ammonites were used by members of the Tifalmin tribe in the Upper Sepik River as charms to help with farming and hunting. In the Harz Mountains of Germany, farmers used "dragonstones" to cure their animals. They believed that putting a dragonstone in a milk pail would induce cows that had stopped lactating to make milk again. In Scotland, ammonites were called "crampstones" and were thought to cure cramps in animals. Their livestock were washed with water containing a "crampstone" to relieve their pain. In Cornwall in southwestern England, they were called "snakestones." Richard Carew wrote in 1620 in his *Survey of Cornwall* that "Beasts which are stung, being given to drink of the water wherein this stone has been soaked, with therethrough recover."

In the Rocky Mountains, ammonites developed a unique mode of preservation that has imbued them with beneficent magical connotations in Blackfoot spirituality. *Iniskim*, literally "Buffalo Stones," as they are known in the Blackfoot community, have been traditionally taken on hunts as amulets for the successful capture of bison (Figure 1.2b). Today, these uniquely preserved ammonites are also known as the highly valuable semiprecious gemstone "ammolite." Ammolite-bearing ammonites generally belong to the genera *Placenticeras* or *Baculites*, which dominated the seaways that flooded the Great Plains during the Cretaceous. They are highly iridescent due to differential erosion over the nacreous (pearly) layer of shell. The differences in thickness of the eroded mother-of-pearl enable the full spectrum of the rainbow to emerge from the ammonite fossils.

The Greeks and Romans saw ammonites as sacred symbols representing their horned god Jupiter, whose Egyptian equivalent, Ammon, sported a pair of ram's horns shaped much like an ammonite fossil. In fact, the name "ammonite" came from the Latin *Cornu ammonis* ("horn of Ammon"). They were also used as talismans for protection from snakebites, and as a cure for blindness, infertility, and impotence. The Romans prized "gold" ammonites from Ethiopia (actually, fossils replaced by pyrite or "fool's gold") and placed them under their pillows so the dreamer could predict the future.

In medieval England, Jurassic ammonites gained the common name "snakestones" (Figure 1.3a), partially due to the legend of St. Hilda, who had considerable cultural influence during the early Middle Ages. In what is probably a rehashing of the legend of St. Patrick, in the late 600s, England and Ireland were in the process of conversion to Christianity. It was said that St. Hilda, a princess-turned-abbess in Whitby, Yorkshire, removed a local snake infestation by miraculously turning them into stone. She allegedly prayed about them so that suddenly their heads fell off. The evidence for this, of course, was the presence of coiled rocks, the ammonites, most often

Figure 1.3 (a) A "snakestone," an ammonite whose similarity to a coiled snake has been enhanced with a carved head. Snakestones were commonly made from *Dactylioceras* or *Hildoceras* ammonites. (b) A depiction of St. Hilda with snakes at her feet, from a monument in the graveyard of St. Mary's Church, Whitby, Yorkshire. 1.3C. The coat of arms of the town of Whitby, with the snakestone in the center.

the genus now named *Dactylioceras*. The genus *Hildoceras*, also abundant at Whitby, is named after St. Hilda. Snakestones are featured as accents in gothic architecture, often with the heads of snakes carved onto their apertures.

Sir Walter Scott, in his poem "Marmion," wrote:

> When Whitby's nuns exalting told
> Of thousand snakes, each one

> Was changed into a coil of stone,
> When Holy Hilda pray'd:
> Themselves, without their holy ground
> Their stony folds had often found

There is also the legend of St. Cuthbert, a seventh-century monk who is said to have cast such a powerful curse on the snakes that it beheaded them before turning them to stone, which is the supposed reason that the common "snakestones" never showed a snake's head.

In Keynsham in Somerset in southern England, there were similar myths about the mysterious snake-like fossils found in the rocks. In this case, it was St. Keyna, a British virgin who lived in snake-infested forests. According to the legend, she prayed and turned the serpents into stone.

So What Is an Ammonite?

Ammonites are technically the Mesozoic subset of the order of cephalopods called Ammonoidea (ammonoids). All ammonites are ammonoids, but not all ammonoids are ammonites. Though ammonite fossils were in the periphery of human life, across many cultures and for thousands of years, it was not until about a few centuries ago that we determined what animals they represented. Although imported nautilus shells were popular as curious objects in collectors' cabinets in the Renaissance, but it wasn't until 1785 that the chambered *Nautilus* was formally named by the founder of modern classification, Carolus Linnaeus, based on empty shells. But a complete specimen with the soft tissues was not formally described until the 1832 in *Memoir on the Pearly Nautilus*, by the famous British anatomist and paleontologist Richard Owen. It was not until recent *decades* that we have begun to look beyond the veil into how ammonoids and nautiloids lived. Ammonoids were not serpents mystically entrapped by a legendary saint, nor were they huge, faceless snails. Instead, they were an ancient subclass of cephalopods which lived from the Middle Devonian Period, about 390 million years ago, to just beyond the end of the Cretaceous Period, about 64 million years ago, or about 326 million years in duration. Their extinction was ultimately set into motion by the events that also wiped out the dinosaurs.

To most people unfamiliar with them, ammonites appear to be nothing more than giant snail shells. In fact, in Germany, locals directing fossil-hunting tourists simply call them *große Schnecke*, or "big snails." However, looking inside ammonites reveals a much more complicated story. Unlike snail shells, ammonite shells are chambered like a nautilus. Walls of shell seal off sections of the inner shell, allowing the chambers to be filled only with fluid or air. This is how cephalopods control their buoyancy, and it is the telltale sign of a cephalopod shell. The next indicator that ammonites were cephalopods is the thin tube that connects all of these chambers, called a siphuncle (or siphon). In life, this tube was filled with especially salty fluid, which helped it to absorb water from the chambers when ammonites needed to float upwards. The siphuncle, and its fleshy exterior terminus, called the hyponome, still exists today in the chambered *Nautilus* and other cephalopods (Figure 1.4). Because of these features of their shells, we know that ammonoids, the animal counterparts to the iconic ammonite fossil, were the close cousins of octopodes and squid, and the distant cousin of the chambered *Nautilus*, a simpler version of which is the taxon from which all cephalopods derive. We also know that ammonoids exhibit traits not found in any known cephalopods before or after them. We are even beginning to understand how they finally met their end, despite surviving several other mass extinctions, including several that were quantifiably *more* catastrophic than the one that ultimately killed them. Soft-part fossils of the animal inside the shell are still extremely rare, so reconstructing the living animal is a real challenge—yet there are clues.

When thinking of ammonites as living cephalopods and not as spiraled rock curiosities, we must consider all of the recent discoveries in both the soft and hard tissue of ammonites to depict them as complicated, forceful players in their ancient ecosystems. The size of the animal's living chamber can tell us much about the animal's buoyancy—did it float near the surface, or did it practically touch the seafloor? Pieces of the jaws and beaks of ammonites can help us describe what a specific ammonite ate. From understanding how the animal hunted food, we can better understand how the animal evolved to move. The manner in which an ammonoid moved relates to the most likely diet of each species, and where the fossils were found interact in a unique way in each species. Members of the ammonite genus known as *Didymoceras* exhibit both a large living chamber and sizeable aptychi (bottom jaw attachment pieces) and in this book, they are therefore reconstructed as more muscular, more octopus-like, if you will, than the more gracile *Sciponoceras*, whose delicate mouthparts indicate a life of passively ingesting microscopic planktonic crustaceans. Other parts of the shell indicate that many ammonoids some had neutral buoyancy, neither sinking nor floating. These ammonoids likely floated just inches above the sea bottom, waiting for food to come to them. Some of them lived around methane seeps—oases of biodiversity that teemed with life in select locations along the otherwise treacherous, low-oxygen bottom of the Western Interior Seaway—and they would not likely have waited long.

(a)

(b)

(c)

Figure 1.4 **(a) An ammonite reconstruction (*Harpoceras falsifer*) next to (b) a living chambered nautilus (*Nautilus pompilius*) and (c) a blue-ringed octopus (*Hapalochlaena lunulata*). While the external shells of the nautilus and ammonite make them appear to be the most similar of the three at first glance, the low arm count and large eyes in muscular, raised eye capsules actually show greater similarity between the ammonite and the octopus.**

It bears repeating that ammonoids are more closely related to squid than they are to the chambered *Nautilus* (Figure 1.5). The external shell is only a superficial similarity, as nearly all cephalopods have some form of a shell, many even with chambers, and ammonoid shells differ considerably from the shell of a nautilus. Like their octopus and squid cousins, it is believed that ammonites boasted similarly prodigious intelligence, and their

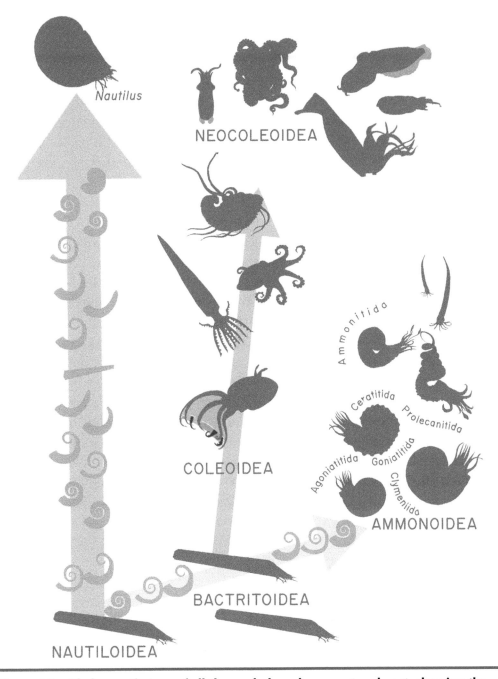

Figure 1.5 Phylogenetic tree of all the cephalopods present and past, showing the taxonomic distance between all of the ectocochleate cephalopods: ammonoids, nautiloids, and argonaut octopodes. Based on Hoffmann et al. 2022.

high degree of specialized biodiversity ensured they were the unmatched geniuses in all of their ecological niches. However, all cephalopods are derived from a nautiloid ancestor. Throughout the Cambrian and Ordovician, nautiloids flourished, rivaling trilobites as the dominant complex animal. They started as straight-shelled, or orthocone, forms, which evolved into the well-known planispiral shape after curling over into several renditions of an open-coil spiraling conical shape. They eventually straightened back out, and all of this restructuring of the shell resulted in myriad sizes, shapes, and specializations for nautiloids. Shells were often thick, and they were used at least as much for flotation as they were as a defense of the vital organs.

Ammonites are most commonly found in a few specific types of rock. The most common rocks with ammonites are limestones, sandstones, and especially shales. Ammonoids in limestone probably lived slow, calm lives on a coral reef, which is more common in the Paleozoic than the Mesozoic. These animals often include goniatites, the ancestors of ammonites. Sandstones indicate shallow seas and often the intertidal or beach zone, and can indicate a shell that was washed in from elsewhere and deposited near shore after the ammonite died. A shale matrix implies that an ammonite lived either in or above depths of anywhere from 40 to over 150 meters. We then use other fossils in the environment to determine the depth at which the ammonite in question lived. Some of these ammonites may have adapted for vertical migrations, rising to the surface nightly or some similar cycle. Isotopic evidence shows us that some did this seasonally. Operating a chambered shell similarly, in some ways, to the chambered nautilus suggests that many ammonoids likely migrated vertically throughout the day, or sped through the open sea the way that schools of tuna do today.

What Is a Fossil?

Despite the centuries of mythical and mystical beliefs about ammonites, early scientists began to examine the mysterious coiled shells and discover their true nature. In those days, the nature of fossils was very controversial. The presence of fossils seemed to undermine the Great Chain of Being, a world-view that suggested a benevolent god could not let species go extinct. Some thought they were caused by supernatural forces (*vis plastica*, or "plastic forces" in Latin), either planted in the rocks by the Devil to shake our faith, or "sports of nature" (*lusus naturae* in Latin). But there were some scholars who correctly realized that fossils are the remains of once-living organisms.

Five centuries after Su Sung suggested that ammonites were an ancient conch-like creature, Danish physician and naturalist Niels Steensen (known to us by his Latinized name, Nicholas Steno) confirmed that notion in 1669, and laid the foundations for telling time in geology. Unfortunately, he never followed up on his pioneering ideas. Around 1500, Leonardo Da Vinci drew sketches of ammonites (Figure 1.6a) in his secret notebooks, but they were not published until recently. He correctly realized that if fossils were found in the rocks of the Apennine Mountains behind Florence, and they were in life position and delicately preserved, they were not tossed there by Noah's flood but had been buried by a gentle coating of mud and sand—so the rocks they were found in were not Noah's flood deposits, but ancient relics of a prehistoric past.

The legendary naturalist Robert Hooke (1635–1703) was also one of these pioneers. Today he is famous for his physics experiments and for the understanding of the physics of springs (now known as Hooke's law), but he was a true Renaissance man, with contributions in many fields of natural science. He pioneered the use of an early microscope to be one of the first to document the existence of cells. He also wrestled with the vexing issue of the nature of fossils. Hooke pointed out that fossil shells of familiar clams and snails, as well as the mysterious "snakestones," must have once been the shells of animals that had become buried in sediment and then later turned into stone. In 1705, his posthumously published works included an illustration (Figure 1.6b) which showed the shells of several kinds of ammonites, as well as oysters and snails, interpreted as shells of once-living organisms, not magical talismans.

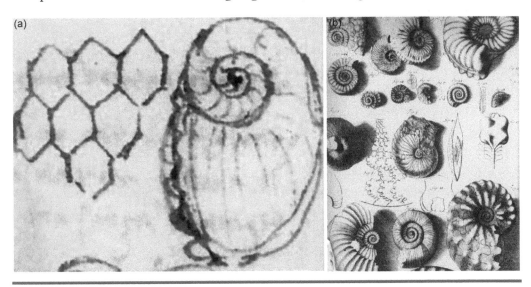

Figure 1.6 **(a) Da Vinci's sketch of an ammonite from his unpublished notebooks. (b) Hooke's (1703) posthumous illustration of ammonite fossils.**

Ammonites and Geologic Time

The practical importance of ammonites was not appreciated until the late 1700s. A humble canal engineer named William Smith (1769–1839) (Figure 1.7a) was hired to survey the route of major canal excavations in southwestern England in 1787. The Industrial Revolution was in full swing, and the great factories needed a cheap means to transport heavy loads of coal from the mines to their furnaces that powered them. Horses and wagons could not carry it efficiently on those rutted bad roads, so canal boats and canals were the solution. Within just a few decades, though, railroads made the canals obsolete. Smith, a keen observer and avid fossil collector, saw opportunity as the canals were dug to get the world's first good look at the fresh bedrock below the normally lushly vegetated English landscape, where bare rock exposures are few. Soon, he could clearly identify the different Jurassic rock formations in the area where he was surveying the Somerset Coal Canal south of Bath, England. More importantly, he got good at recognizing the characteristic fossils found in each formation. He discovered that fossils changed through time, and that you could tell what formation you were in, and therefore what time it was, if you knew the sequence of fossils through time. This is the principle of faunal succession, and it is one of the most important ideas in all of geology, because it is crucial to how we tell time in ancient rocks. This is the basis for an entire specialty known as biostratigraphy, telling the age of rocks by their fossil content.

Smith amazed the rich gentlemen fossil collectors of his society by his ability to tell them exactly what formation each of their fossils came from (Figure 1.7a) and used his ideas to map the formations across southeastern England. Sadly, they treated the humble canal excavator Smith with scorn since he was not an educated wealthy gentleman, and some even stole his ideas and took credit for them. Smith made almost no money from his work, and even ended up in debtors' prison for a while, but eventually, in 1815, he published the first true geologic map, which covered all of England and Wales and parts of Scotland. It is still accurate today at the scale at which it is drawn! The author Simon Winchester called it "the map that changed the world," because a geologic map is fundamental to almost any work in geology today. Smith struggled to make a living as a surveyor and avoid debt for many years, but by the time he died in 1839, he was recognized as "the father of English geology." And recognizing the age of Jurassic ammonites was one of the early keys to his discoveries.

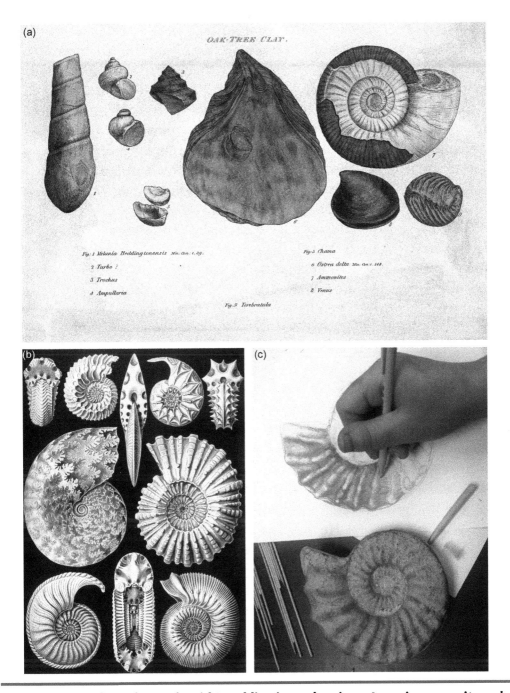

Figure 1.7 **(a) Plate of one of Smith's publications, showing a Jurassic ammonite and other associated Jurassic fossils. (b) Ammonites by Ernst Haeckel. (c) Kate drawing a *Pleuroceras* specimen in Alex's lab.**

Smith's discovery took advantage of the fact that some fossils are particularly good index fossils, that is, good time markers. Index fossils have to be animals that change rapidly through time (so their species are short-lived, and you can subdivide time in finer resolution), are easy to identify from other species, are abundant in most outcrops of the right age and environment, and are geographically widespread, so the same fossil found in several different regions can be correlated together. Ammonites are particularly good in this regard, because they were among the most prolific and rapidly evolving animals in the Mesozoic. They tend to occur in huge numbers in the right environments; because most of them floated over large areas of ocean, breeding rapidly and possibly dispersed by currents, allowing them to become very widespread. Even if they didn't live on the sea bottom with its distinct water depths and environments, many of them floated or swam in the surface waters so they sank to the bottom of nearly every marine setting and were fossilized. Once you are familiar with ammonoids, most are fairly easy to identify. For these reasons, ammonoids are one of the principal index fossils from the Devonian through the end of the Cretaceous.

The idea of faunal succession is so important that inevitably, it became apparent everywhere that fossil successions were discovered. The great French anatomist Baron Georges Cuvier, and the fossil mollusk specialist Alexandre Brongniart first described the rocks around Paris. They noticed a predictable sequence of formations and their fossils and came up with their own version of faunal succession. Some scholars point out that Brongniart visited England in 1805 (during one of the brief periods of peace between France and Britain in the era of the French Revolution, the Reign of Terror, and the Napoleonic Wars). He may have heard of Smith's ideas, then brought them back to Paris and discovered they worked there as well. French geologists say that this is not the case. Whoever was responsible, the idea was ripe for discovery by that time, because so many people were collecting fossils and observing the rocks they came from. In 1840, Cuvier's student, the great French paleontologist Alcide d'Orbigny (1802–1857), documented the stratigraphic paleontology across France and much of Europe, especially using Jurassic ammonites, and in 1849, he described some 18,000 species of fossils. Using ammonites, d'Orbigny was the first to give formal names to many of the subdivisions (stages) of the Mesozoic, including the Toarcian, Callovian, Oxfordian, and Kimmeridgian stages of the Jurassic, and the Aptian, Albian, and Cenomanian stages of the Early Cretaceous.

Meanwhile, naturalists and scholars in other countries were studying their own fossils. In the 1840s and 1850s in Germany, the paleontologist and mineralogist Friedrich Quenstedt (1809–1889) made a special study of Jurassic ammonites, especially those from the Jura Mountains of the western Swiss and French Alps, where the "Jurassic" got its name. He published on hundreds of ammonite fossils from there in his 1883–1885 volumes *Die Ammoniten des Schwäbischen Jura*, although his system of naming forms has caused problems for later ammonite specialists. But his detailed work established how Jurassic ammonites could be used to tell time not only in Europe, but across much of Europe and eventually, the world.

Quenstedt's ideas were further developed by his student Albert Oppel (1831–1865), who studied under him at the University of Tübingen and got his doctorate there in 1853. By 1856, Oppel was publishing a grand synthesis of the Jurassic stratigraphy of England, France, and Germany, using ammonites to correlate these beds across all of Europe. Today, Oppel is famous for developing techniques for finer-scale correlation in fossil sequences. Most scholars before him had been content just to note which fossils were found in which formations. But Oppel carefully plotted the stratigraphic position of his ammonite fossils in the Jurassic rocks of the Alps and elsewhere. Using ammonites, he recognized that one could tell time by documenting the vertical range of a fossil species in a sequence of rocks to form a "range zone." The range zones of each individual species could then be compiled, and when enough of them are compared, you could plot the overlap of the range of one species with the range of another, and each overlapping range zone would give a finer and finer subdivision of geologic time. The principle of overlapping range zones is now the foundation of modern biostratigraphy; it is used to map strata, date rocks, and tell geologic time with high precision, and is the foundation of the entire geologic time scale as we now know it. Ammonite workers continued to make discoveries long after Oppel died prematurely at the age of 35, but we will discuss their contributions elsewhere in this book. The important takeaway is that ammonites were, and continue to be, integral to the story of paleontology as some of the first objects to help us understand fossils as the remains of once-living organisms. Ammonites were also crucial in telling time in rocks and the development of the geologic time scale.

Ammonites and Paleoecology

The rocks in which ammonite fossils are found may be just as important as the fossils themselves in telling how prehistoric organisms looked and behaved. Invertebrates are often intimately tied to their environment in ways where vertebrates may instead be more closely linked to one another. This is especially true of cephalopods, who often only leave behind their shells— and at that, they leave their shells behind in very specific places.

Sedimentary rocks communicate volumes about the environments that formed them. They serve as testaments to water depth and turbidity, temperature, and even the levels of oxygen, salinity, and other chemical factors. Two main categories of fossiliferous sedimentary rocks exist: carbonates and clastics, and both often harbor ancient cephalopods. Carbonate rocks include all limestones, from reef debris piles to lime muds. Clastic sedimentary rocks, for our purposes, consist primarily of sandstones and shales. Sandstones tend to form in nearshore, highly agitated environments, such as the beach and surf zone, or the shallow offshore waters that are disturbed by the waves. Due to the durable nature of quartz sand grains, the rock types which form sand are hard to erode. By contrast, shales are the result of extremely weathered clay particles that settle out where there are almost no currents or waves. They are formed when the most fine-grained clays and silts, easily eroded and often dark in color, wash all the way out to sea and come to rest on deeper parts of the continental shelf, or the abyssal plain, where they are squeezed, by water weight and the rocks piling on top of them, into flat, flaky stones.

Ammonoid fossils are known from limestones and shales, and to a lesser extent, from sandstones. This means that they lived in calm waters in which their thin, delicate shells were unlikely to become damaged. Paleozoic ammonoids may have lived in or around coral reefs, where wave energy and water turbidity were modified by the protrusion of corals into the water column. Mesozoic ammonites are most commonly found in shales and limestones, and their chemical compositions suggest that they often lived at moderate depths. They favored calm waters, almost always a safe distance from shore, in which they were unlikely to get bashed into a rock by a breaking wave.

Knowledge of the environments that form the surrounding matrix of a fossil aids when an animal's soft tissues are completely missing, as is the case for most ammonites. An ammonite shell built for speed but deposited in a dark gray or black shale, such as *Placenticeras*, likely cruised the open ocean at relatively high speeds (for an ammonite), and its soft parts were probably just as streamlined. Some early ammonoids like *Agoniatites* were laterally more robust, and they are found in carbonate deposits with millions of corals, trilobites, and sponges: it may be safe to assume this animal made no real migrations and instead lived off the bottom-dwelling fauna of a single localized area. The drag associated with the whorls of helical *Didymoceras*, another ammonite found primarily in shales, suggests an inability to swim well, but its huge jaw apparatus signals a diet that was not simply filter feeding microorganisms: could it simply have been an ambush predator of the deep? *Nipponites*, whose shells are so bizarre that it seems impossible that the animal had much control over its movement, are often found nearshore, shallow deposits, and are only found on the island of Hokkaido, Japan—except for the few found in Washington State. Is it possible that they were washed in passively? When combined with morphological data, environmental cues from sedimentary geology give these mysterious animals plausibility.

We will explore this topic, and many others, in the chapters to come.

Further Reading

Berry, W.B.N. (1987). *Growth of a Prehistoric Time Scale* (2nd ed.) Blackwell Scientific Publications, New York.

Faul, H., Faul, C. (1983). *It Began with a Stone*. Wiley, New York.

Prothero, D.R. (1990). *Interpreting the Stratigraphic Record*. Freeman, New York.

Rudwick, M.J.S. (1972). *The Meaning of Fossils: Episodes in the History of Palaeontology*. Macdonald, London.

Chapter 2

Anatomy, Growth, and Geometry

Ammonoid Anatomy 101

Ammonoid fossils are commonly thought of as heavy, solid rocks, but from what we understand of the once-living animals, they could not be further from this description (Figure 2.1). For one thing, ammonoid shells were incredibly lightweight. Like their distant cousins, the chambered nautilus, ammonoid shells also had chambers. The hollow upper portion of the shell, known as the phragmocone, was divided into small, sealed chambers. (The phragmocone includes all chambers except the body chamber, in which the animal's soft tissue is housed.) The chambers were connected only by a tube of soft tissue, called the siphuncle, with which they transported water and gasses in and out of the chambers. In this way, ammonoids controlled their position in the water column. Like the other living cephalopods, ammonoids had a nozzle-like opening to the mantle cavity known as the hyponome, by which they could jet propel their bodies out of danger.

Ammonoids differed from nautiloid shells in several key aspects. They shifted their siphuncle away from the center of the chamber, where it is found in nautiloids, toward the ventral, or outside edge. While the structures within the shell that support the fleshy siphuncle in nautilus face backwards (Figure 2.2), ammonoids reversed these straw-like protrusions, called septal necks, apparently positioning them to withstand the brunt of the force that occurred during a sudden expulsion of fluid from the chambers. Many early ammonoids were not built for speed, but appeared so by comparison

DOI: 10.1201/9781003288299-2

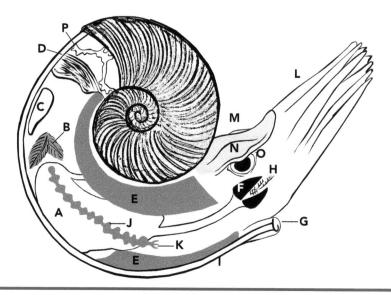

Figure 2.1 **The basic anatomy of an ammonoid. (a) Stomach, (b) gills, (c) repro-ductive organ, (d) unknown laminated structure, (e) retractor muscles, (f) beak, (g) siphon, (h) radula, (i) siphuncle, (j) intestine, (k) anal gland, (l) arms, (m) cephalic mass, (n) eye capsule, (o) eye, and (p) last septum.**

as nautiloid cousins plodded slowly. This rapidity carried over to their lifes-pans. As the early ocean filled with free-swimming predators, ammonoids began to live rapidly sped-up lives. Estimated shell growth rates indicate that ammonoid life expectancies averaged only a fraction of that of the living *Nautilus*. Their egg broods had higher yields, smaller hatchlings, and ammo-noids reached maturity much sooner than their nautiloid cousins.

It is useful to think of ammonoid shells as long cones which have been curled up around themselves in a spiral. As with all mollusks, ammonoid shells have an apex which is the beginning point at which their shell began grow-ing in the first part of their lifecycle. A *whorl* is a full revolution of growth around the axis of coiling. Shells with more whorl overlap are *involute*, and shells with less overlap between whorls are *evolute*. Many "normal," or *plani-spiral*, ammonoid and nautiloid shells have a moderate amount of overlap, and we call them *convolute*. Biologically, the apex is actually a nearly-microscopic coil called the *umbilicus*, which represents the embryonic growth stage of the shell. The first chamber, in which the embryonic ammonoid lives, is called the *protoconch*. The umbilicus is at the middle or innermost portion of the shell in more mature shell, as all new shell growth occurs around it. Ammonoids grew by adding new chambers and new whorls around their coiling axis. A whorl is a full revolution, or 360°, of growth. The opening of the shell, called the aper-ture, represents the latest part of the shell to have grown.

Figure 2.2 **Nautilus shell in cross section. Note the central siphuncle and back-facing septal necks.**

In the time that ammonoids lived on Earth, roughly 50 shell shapes evolved (Figure 2.3). Some very evolute planispirals evolved several times, with whorls so loosely coiled that they separated. The tarphycones opened up slowly and were followed by open-coiled gyrocones. Eventually, long, straight shafts were added, and in the Middle Jurassic, the first ancylocones appeared. Helical forms emerged, including the turricone, emperocone, and nostocone. In the Late Cretaceous, truly outlandish shapes appeared, including the vermiticone, the polyptychocone, and even the heterotrianglicone. However, the simplest forms proved the most enduring, and plain old planispirals persisted until the total extinction of ammonoids in the early Cenozoic.

Jaws

Aptychi are double-valved calcitic structures found in association with ammonoids. They represent the outermost portion of the lower beak of the living animal. Anaptychi, which are less common, are interpreted to have been part of the upper beak. Each valve represents either the left or right

Figure 2.3 There are approximately 50 shapes in which ammonoid shells grew, but even planispirals had a surprising amount of variation.

side of the lower or upper jaw. For ammonoids which were encased in sediment shortly after death, it is not the rarest thing to find aptychi inside the living chamber. Paleontologists call this *aptychus-in-situ*.

The bifid mystery structures known as aptychi were originally thought by paleontologists to merely be fossilized clams occurring incidentally in the same outcrops, and sometimes, inside the shells of ammonoids. It eventually became apparent that the "clam," named *Trigonellites latus*, was no bivalve, and that it was definitely part of the ammonoid body. They were then considered opercula to the opening of the shell. For many years, paleontologists argued that in moments of peril, an ammonite could simply pull its entire soft body into the living chamber and essentially shut the door via this apparent operculum.

However, these interpretations have been debunked, and the consensus surrounding the aptychi is that they acted as support structures to the animal's beak, or even the beak mandible itself. All known cephalopods have a beak, and ammonoids were no exception. It is somewhat, but not exceedingly rare to find an *aptychus-in-situ*, or in the place it would have been located during life. Aptychi belong under or surrounding the animal's radula, a tongue-like structure in mollusks that looks a bit like a feather duster made of teeth.

Aptychi are important when determining how ammonoid species looked and behaved because they give us clues when trying to understand what an animal ate. Feeding behaviors can indicate anything from environmental information to various aspects of an animal's appearance or mobility. Small aptychi often accompany a delicate radula and are thought to be evidence of a filter-feeding ammonoid. A large, thick aptychus suggests the ammonoid ate large prey (Figure 2.4).

Cornaptychus is common among the older and more delicate forms of aptychi. These ammonoids probably filter-fed or scavenged carrion. Belonging primarily to hildoceratids, these were specialized for foods that were not difficult to break apart. Because they belonged to a common Jurassic family, *Cornaptychus* is the most common constituent of the aptychus beds in Europe's Ammonitico Rosso (see Chapter 7).

Granulaptychus is a very powerful aptychus, with each valve covering half the area of the shell aperture. It used to be interpreted as an operculum, or cover, for the shell opening if an ammonoid decided to hide its whole body inside its shell. This interpretation is now considered obsolete.

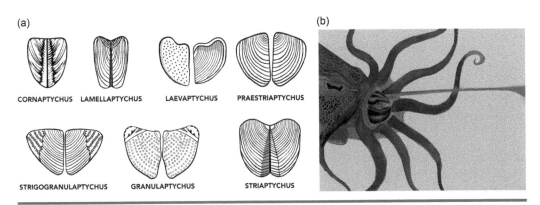

Figure 2.4 (a) Aptychi of ammonoids and (b) an ammonoid eating.

Striaptychus were somewhat varied structures, united only by the character-istic commissure down the middle. *Striaptychus* is found only in heteromorph ammonoids, such as the ones on the cover of this book, and can be seen, depending on species, as either macrophagous (eating large prey) or passive microphagous (passively consuming small prey, like plankton) mandibles. In the case of *Didymoceras* and its closest relatives, these aptychi seem to have been used on large prey items, because they have narrowed and developed a sharp point. For *Nipponites* and smaller baculitids, they were probably used to open and close the gastric cavity when the animal had had enough plankton.

Radula

The radulae are intricate probing structures in the mouths of mollusks. They are covered by hundreds of tiny teeth. In snails, the ribbon of teeth in the radula is used to rasp algae off of a rock, or to rasp holes in the shell of prey animals. In ammonoids, these delicate structures are almost never preserved. The radula (Figure 2.5) can provide similar information about

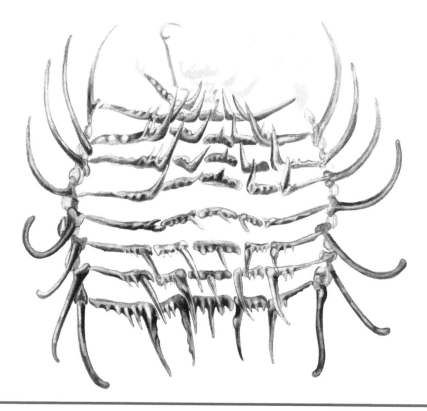

Figure 2.5 **Radula of a baculitid. Redrawn from Kruta et al., 2010.**

the behavior and soft-part morphologies of ammonoids via what the animals ate. A small baculitid ammonite with a radula apparently adapted for filter-feeding was found, and its stomach contents were indeed dominated by microscopic, free-floating crustaceans.

Soft Tissues

Eyes

There is precious little in the way of ammonoid soft-tissue fossils, but from what has been ventured, it is believed that the eyes of these cephalopods were much closer in both appearance and function to squids and octopodes than they were to nautiloids. It is possible that early ammonoids shared more superficial features with nautiloids, and that these features eventually gave way to more neocoleoid-like appearances and behaviors, though many were likely impeded by their enormous gas-filled shells.

Several specimens of the Cretaceous straight-shelled ammonite *Sciponoceras*, found in 2012 by Christian Klug and his colleagues, show an eye preserved in the dorsal view that appears much closer in appearance to the eye of a cuttlefish or squid. It was long and sharply honed, likely encapsulated in a long, muscular dome, and very different from the blurry, primitive eye of a nautilus. In fact, it is now believed that the "pinhole-camera eye" of living nautiluses, featuring a small, inflexible pupil and relatively poor vision, evolved somewhat recently and is unique to recent nautiluses. For an ammonoid to have eyes like a squid, the animal's perception of color must have been considerable, and we might consider a modern octopus or squid eye to be the most comparable example of the ammonite eye (Figure 2.6).

Octopodes and squid have adapted unusual methods (Figure 2.7) to see more colors than nearly any other animal on Earth. The difference between their vision and ours is that they aren't able to see all of those colors at once.

Figure 2.6   **Presumed eye capsules and soft tissue of *Sciponoceras*. Redrawn from Klug et al., 2012.**

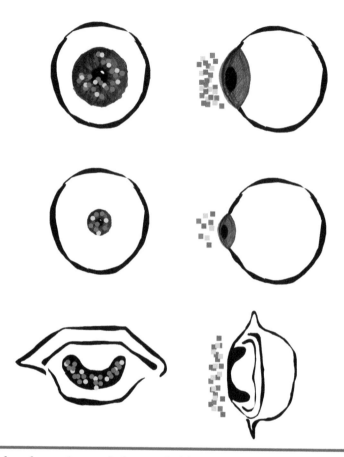

Figure 2.7 **Color absorption and chromatic blurring in cephalopod pupil shapes. The top image is a hypothetical large round pupil; the second is hypothetical small round pupil; the bottom one is a U-shaped coleoid pupil. The large round eye and the U-shaped eye can take in about the same amount of color. The small round eye can take in fewer colors. Redrawn from Stubbs and Stubbs, 2016.**

Color vision, as we understand it in most animals, depends on one thing: microstructures in the iris and pupil called cones that accept and reject certain wavelengths of light. Dogs have two different types of cones and can probably see yellows and greens. Humans generally have three cones, that is to say that we are trichromatic, and we see an average of about nine million different colors. An order of magnitude above that are the tetrachromats, rarified four-coned retinas found in birds, some fish and perhaps a handful of chromatically gifted humans, all of whom are able to see about ninety million to a hundred million colors.

An Australian reef cuttlefish can see all the colors a tetrachromat can, and maybe more: it sees, possibly, close to as many colors as a mantis shrimp. In fact, octopodes and squid camouflaging behaviors are heavily dependent on the tens of millions of unfathomable colors that these animals can see. The caveat? Cephalopods have not a single cone to their name.

Sir Isaac Newton infamously drove a needle into his own eye for the sheer purpose of documenting the explosion of colors that resulted for an ephemeral moment. The way a cuttlefish sees is somewhat similar. The elongate W-shaped pupil is encapsulated within a muscular and somewhat raised eye. By pushing the muscles of the eye inward in one place or another, the animal configures its own color mapping system as it goes. The nuance and intensity with which cuttlefish perceive color, even through salt water, is something of a miracle. Perhaps among the last ammonoid eyes left on Earth, this eye appears—at least superficially—much more similar to that of a cuttlefish than to the primitive eye of a nautilus. It is possible that colors, to a light-dwelling ammonoid, were nearly as profound as what our cuttlefish and octopodes see today.

The father-and-son research team Alexander and Christopher Stubbs published an influential interpretation of cephalopod color perception in 2016. The size and shape of the pupil, they argued, affected the degree of "chromatic blurring," that is, the amount of colors filled in by the brain from minimal visual stimulus. Due to their pupils, which change in size, shape, and pressure, cuttlefish are able to at least perceive wavelengths of color, regardless of whether or not they actually see them as we do. The fact that they may not see colors makes their remarkable camouflaging abilities all the more incredible.

Arms

Arms and tentacles do not refer to the same body parts for cephalopods. Tentacles are retractable, and arms are not. An octopus has eight arms and zero tentacles, but a squid has eight arms and two tentacles. In octopuses and squid, the arms may be lined with suckers, serrations, hooks, or cirri—long, fleshy sensory organs. Nautiluses simply have over 90 large cirri which act as arms. Most paleontologists have assumed that ammonoids had some number of arms for decades. However, the debate over what kinds of arms ammonoids had, if any, has also persisted. Given that ammonoids are much closer cousins to octopodes than to nautiluses, a low arm count was expected by most researchers. Tentacles were even less expected. Some researchers like Dolf Seilacher argued that ammonoids may have had no arms or tentacles whatsoever. However, the *Sciponoceras* fossil discovered in 2012 (Figure 2.6) also exhibited a low arm count. This confirmed what many paleontologists expected for years. Still, it was impossible to tell from that particular *Sciponoceras* fossil how long its arms may have actually been. Its arms were found entangled by the hooks of a more powerful cephalopod, a Cretaceous squid relative called a belemnite. While all ammonoids likely possessed a similar arm count to *Sciponoceras*, the full diversity of sizes and shapes is still a bit of a mystery.

Mantle

This structure (Figure 2.1) is present in all mollusks, from quahog clams to giant Pacific octopodes. This fleshy mass has several diverse functions in mollusks, and cephalopods are no exception. For ammonoids, it is sometimes interpreted as covering and protecting the animal's digestive and reproductive organs. It also interfaced in some capacity with the back of the living chamber, creating new chamber walls. It is still unresolved whether this is due to fluids interacting at the back of the mantle or if the mantle's texture was simply just frilly. Paleontologists have long debated how much of the animal's body was actually covered by mantle. In modern nautiluses, the mantle is inverted and lies underneath the gastrointestinal tract.

Jet Propulsion

While ammonoids used their siphon to move backward and to adjust their buoyancy (as does the chambered nautilus), they did not "jet propel" the way the octopuses do. Most of them would have been relatively slow swimmers when compared to squid or octopodes, but a select few may have been fast backward swimmers compared to *Nautilus*.

The siphuncle (Figure 2.8) is the tube of tissue that connects each of the shell's empty chambers. In nautiluses, this tube is very easy to pick out, as the septal necks run directly back through the chambers down the center of each one (Figure 2.2). In ammonoids, the siphuncle is normally positioned at the ventral shell wall, making it a little more difficult to see unless the external shell layers are removed. The fleshy endpoint of this tube, called the hyponome, occurs under the eyes and beaks of the ammonoid, near its arms. The hyponome is extremely important for jet propulsion. Nearly all modern cephalopods have a visible hypostome: it is the open nozzle-shaped tube through which they push jets of water from their mantle cavity before shooting backward.

The siphuncle is supported by septal necks, residual shell material that grows around it a few millimeters out of each septum. In nautiloids and very

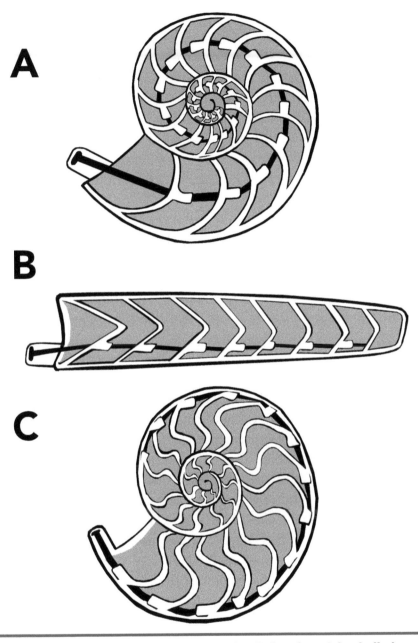

Figure 2.8 (a-b) Siphuncle and septal necks of coiled and straight-shelled nautiloids and (c) siphuncle and septal necks of a bactritoid, which does not run through the center of the septum, and (c) siphuncle and septal necks of an ammonoid, running through the outer (ventral) edge of the septum.

primitive ammonoids, the septal necks are said to be retrochoanitic, meaning that they point back into the shell. However, ammonoids eventually face their septal necks forward.

With the exception of the most primitive of ammonoids, who were still in transition, the direction of the septal necks normally point in the same direction as the convex side of the septae. When ammonoids finally pointed their septal necks out, to face the same direction as the hyponome, backward propulsion would have become more powerful, as the siphuncle would be much better supported for faster backward movement than that of a nautilus.

The Living Chamber

The final chamber of the shell in a nautiloid or ammonoid, the one that would contain the animal's soft body, is called the living chamber. When reconstructing a cephalopod whose soft parts have not been preserved (which is the case for nearly every cephalopod in the fossil record), living chambers provide a remarkable amount of information. Living chambers are nothing more than empty spaces, and because of sediment, even secretion points can generally not be fleshed out for ammonoids. However, the size and shape of the living chamber communicate a lot, particularly when analyzed relative to the entire shell.

For most ammonoids, the living chamber is not measured as a percentage of the shell, but rather by the angle made by the distance between the aperture and the living chamber's back wall. There are only a handful of angles that could occur in adult shells. This is because the length of the chamber relative to the phragmocone alters the center of gravity for the floating shell. As the animal grew, the orientation of the shell probably changed. This was particularly visible in the case of Mesozoic heteromorphs, such as *Didymoceras* and *Nipponites*, and very rarely in Paleozoic goniatitids. The greatest number of empty chambers possible would always be oriented upward, and aperture orientation acted as a real-time function of this.

The majority of ammonoids were heavily reliant on the buoyancy controls made possible by the phragmocone, but a minority of forms got rid of them almost altogether, filling their shells not with air, but with flesh. There are several categories for the length of the living chamber (Figure 2.9), which are designated by the angle it creates (its length, in essence) around the spiral axis of coiling. A brevidome living chamber does not wrap around the shell

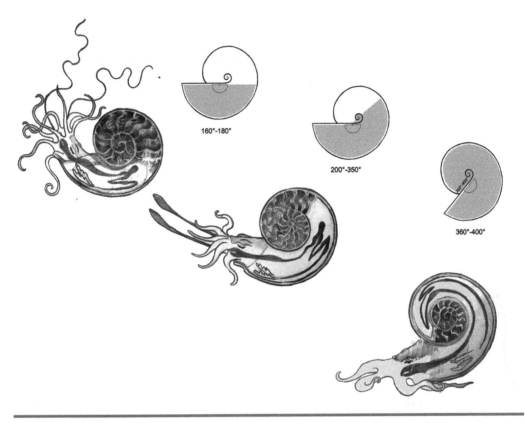

160°-180°

200°-350°

360°-400°

Figure 2.9 **Living chambers of ammonoids: (a) Brevidome, (b) Mesodome, and (c) Longidome.**

axis past 180° and will not be below around 160°. This ratio is in line with modern nautiluses, which are heavily dependent on buoyancy, and this is the model favored by goniatitids and Early Jurassic ammonoids. Uncoiled and rearranged as planispirals, a large number of Cretaceous heteromorphs would fit into this category. The mesodome describes an angle between the aperture and final septa between 260° and 300°. These ammonoids would be less dependent on buoyancy than their more buoyant relatives. The longidome ammonoids have terminal septae within 350°–400° of the aperture's profile. Assuming that a full circle is 360°, these ammonoids have few, if any, buoyancy chambers, and probably lived on the seafloor. Longidome ammonoids tend to be the ones with more involute shells. The length of the body chamber has been used, historically, to determine a given species' position in the water column: smaller body chambers probably had a slight upward orientation and positive buoyancy. Forms with longer body chambers would often be heavier, and less free to float, so they likely stayed closer to the sea bottom. Before modern technologies such as isotope geochemistry (which will be discussed later), this was the primary way to determine where an ammonoid lived relative to both the seafloor and the surface. Through creating models both in real life and in computer simulations, the body chamber continues to be a valuable paleoecological indicator.

This short review of basic ammonoid anatomy and shell geometry lays the groundwork for the description of specific groups of ammonoids in the chapters that follow.

Further Reading

Burnaby, T. (1966). Allometric Growth of Ammonoid Shells: A Generalization of the Logarithmic Spiral. *Nature*, 209, 904–906. https://doi.org/10.1038/209904b0.

Klug, C., Korn, D., De Baets, K., Kruta, I., Mapes, R.H., eds. (2015). *Ammonoid Paleobiology: From Anatomy to Ecology*. Springer, Berlin.

Klug, C., Riegraf, W., Lehmann, J. (2012). Soft-Part Preservation in Heteromorph Ammonites from the Cenomanian-Turonian Boundary Event (OAE 2) in North-West Germany. *Palaeontology*, 55(6), 1307–1331. doi: 10.1111/j.1475–4983.2012.01196.

Parent, H., Westermann, G.E.G., Chamberlain, J.A. (2014). Ammonite Aptychi: Functions and Role in Propulsion. *Geobios*, 47(1–2), 45–55.

Chapter 3

Spirals and Squiggles: The Math of Ammonites

As we discussed in Chapter 2, ammonite shells (and most other cephalopod and gastropod shells) can be thought of as a cone that has been rolled up on itself. Like other seashells both present and past, many ammonoids and nautilus shells exhibit the Golden Ratio, meaning that the spiral in which they grow follows the Fibonacci sequence. The Fibonacci sequence takes the second number in addition equation plus the sum to arrive at the next equation in sequence. It goes:

$$0 + 0 = 0,$$
$$0 + 1 = 1,$$
$$1 + 1 = 2,$$
$$1 + 2 = 3,$$
$$2 + 3 = 5,$$
$$3 + 5 = 8,$$
$$5 + 8 = 13,$$

and so on, approaching infinity. When plotted on a coordinate plane, the square units of each point in the spiral's growth reflect this Golden Ratio. The equation for this is

$$F_n = F_{n-1} + F_{n-2},$$

in which F_n = the next Fibonacci number.

DOI: 10.1201/9781003288299-3

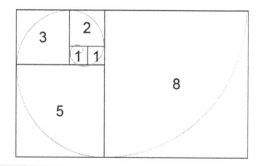

Figure 3.1 **The Fibonacci spiral overlain on a grid of ever-growing squares. The lengths of the squares' sides, in a given unit of measurement, are the Fibonnaci numbers.**

The Fibonacci spiral (Figure 3.1) is often shown overlain on a series of nested squares, which get larger as the spiral progresses outwards. The lengths of the sides of the squares are determined by the numbers of the Fibonacci sequence. The first two squares each have side lengths of one unit. The third square has a length of two units. The fourth has a length of three units and so on. In Figure 3.1, we see that the spiral forms a perfect quarter-circle within each square.

Phi

The "Golden Ratio" owes its appearance of near-perfect organization to a specific value. Like pi, this somewhat obscure irrational number (a decimal whose places are infinite) is commonly represented by a Greek letter. Φ, or phi, is roughly equal to 1.618. When Φ is converted to a proper fraction, its quotient is exactly one less than Φ: 0.618. When it is squared, its product is one greater: 2.618. It can be expressed by the equations:

$$1/\Phi = \Phi - 1 \quad \text{and} \quad \Phi^2 = \Phi + 1$$

$$1/1.618 = 0.6.18 \qquad (1.618)^2 = 2.618$$

The reason that few ammonoid shells deviated from this ratio is that with each place the spiral intersects the square pattern, a point along the curved path of the spiral is exactly 1.618 as far from the shell's apex, or beginning, as it was at the previous meeting of spiral and square. Even the outlandish shell shape known, the tangled *Nipponites mirabilis*, clearly follows the Fibonacci series when viewed in cross section. The shell, which baffled

(a)

(b)

Figure 3.2 **(a)** *Phylloceras,* **an involute ammonoid and (b)** *Lytoceras,* **an evolute ammonoid.**

paleontologists for decades (and still does), is not really a mathematical departure from the Golden Ratio.

The basic shell shape of most ammonoids (and nautiloids) is called a planispiral, or spiraling in a single plane, a shape often used to describe mollusk shells. Planispiral shells are tube-shaped shells that coil along one axis in a flat plane, with the diameter of the tube increasing directly with distance from the shell apex. This general description can result in several different manifestations, each adapted for a different part of the ocean as well as a different behavioral niche. The degree to which a planispiral shell curls over on itself is another component, because in this model it is possible for the tube to overlap itself on each subsequent whorl. A shell that does not overlap whorls at all is evolute, and if it is heavily overlapped, we call it involute (Figure 3.2). Evolute and involute are, of course, relative terms that exist on a gradient. The majority of ammonoids were moderately overlapped, like most nautiluses, with each new arc of growth overlapping the previous whorl only somewhat. A shell in which each new whorl completely envelops the one preceding it, such that the actual spiral shape is hidden from sight, is completely involute. It is possible for the juvenile stage of a shell to be evolute or convolute, and the final whorls to be involute, as is the case for the Cretaceous ammonoid *Placenticeras.*

In 1965, the paleontologist David Raup devised a system to measure the degree of to which a shell is involute or evolute (Figure 3.3). His system

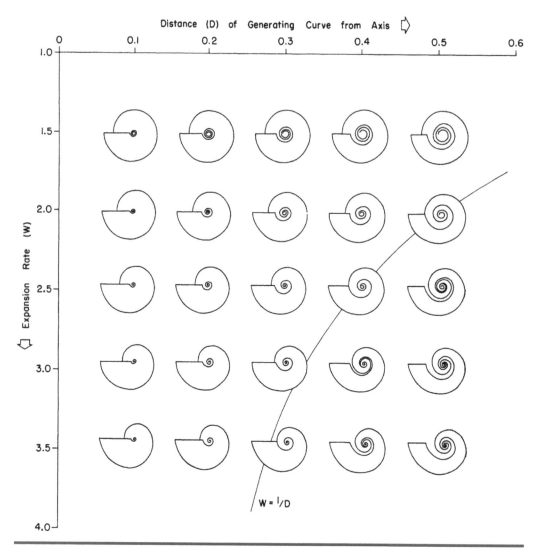

Figure 3.3 **Raup's metric for shell coiling. *W* is the rate of change in width, or whorl expansion rate, *D* is the distance from the generating curve, or how much new shell length has been added to the coil. *S* is the length-to-width ratio, as well as shape, of the aperture at a given point in ontogeny.**

measured for *W*, the expansion rate of the whorl cross section, versus *D*, the distance of the generating curve (position of the aperture relative to the umbilicus). The *W* and *D* values gave way to *S*, which describes how laterally compressed or robust an ammonoid shell is in cross section. *S* values can change over time, making cross-section diagrams (Figure 3.4) very useful.

The oxycone, or the discus-shaped shell with a sharp-edged ventral keel, is a laterally thin shell with a low *S* value. These shells famously belonged

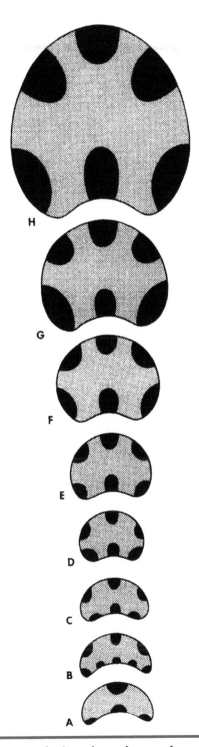

Figure 3.4   **The aperture can and often does change shape over the lifecycle of the animal.**

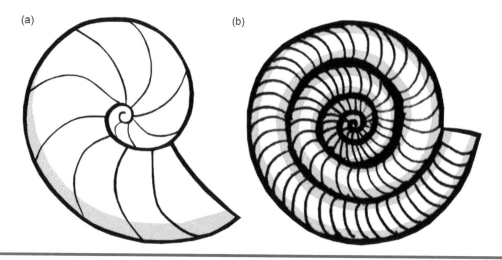

Figure 3.5   (a) A Fibonacci shell. (b) An Archimedean shell.

to *Placenticeras, Sphenodiscus,* and other ammonoids who were associated with darting through the open sea. Cadicones were their opposite; they were extremely wide, fat, robust shells with high *S* values and wide apertures (Figure 2.3).

While the Fibonacci spiral is the most common in ammonoids, some shells followed other mathematical spiral patterns. A shell shape known as "serpenticone" (think of the "snakestones" from England—Figure 1.3) grew not in the Fibonacci spiral, but the Archimedean spiral (Figure 3.5). This pattern is just as predictable as the Fibonacci sequence, and its basic equation is

$$r = a\theta$$

in which *r* is the radius at a given point in the spiral from its apex, and θ is the angle at which the point is located in a coordinate plane; *a* is our constant. In an Archimedean spiral, new whorls are consistently moving away from the previous revolutions at a radian value of $2\pi b$. In this context, *b* is concerned with the degree of tightness between whorls.

While a consistent pattern emerges here, a more elaborate representation of this sequence exists. Archimedes included a sequence, not of numbers, but of equations, for this much more nuanced spiral in his classic book *On Spirals*. The ammonoids that follow this sequence instead of the Fibonacci sequence are known as serpenticones, and they include some of the Paleozoic clymeniids, and Mesozoic ammonites *Dactylioceras* and *Speetoniceras*. Ammonoids are isometric in theory, but not necessarily in practice. Despite the logarithmic pattern ammonoid shells follow, they actually change their chamber length periodically during ontogeny. This is partly due to the fact that neither the ventral edge nor the umbilical seam—the spiral formed at the overlap of the whorls—usually follow the logarithmic sequence, and usually the Fibonacci spiral is only visible in cross section.

Fractals

Imagine we have three values. Value A represents the total value. Value B and Value C together add up to the whole, but Value B is bigger. Numbers are considered to be within the Golden Ratio when the ratio of A:B is equal to the ratio of B:C. Although A:B is a larger total than B:C, proportionally, they are the same. The replication of perfect miniatures nested inside (or simultaneously dividing and comprising) the whole invokes a second mathematical paradigm: the fractal. Fibonacci sequences are, by virtue of their internal replication, fractals.

Real fractals are physically impossible in three dimensions. When you zoom into a fractal image, you discover an identical replica of the whole picture. Further zooming in, into infinity, results in the same thing. The constituent parts are identical to the whole they comprise. The classic example of a fractal would be the Koch Snowflake (Figure 3.6). As new points, or folds, are added to the star, they become identical to the larger shape as they continue dividing into infinity. This means that "true fractals" would need to operate outside of three dimensions, and they are measured instead for their ability to fill the surface they are on with either positive or negative space. The fractal dimension, therefore, is between 1 and 2.

Putting a number to how complex an ammonite suture can be done by taking the fractal dimension of the suture pattern. In 1977, the

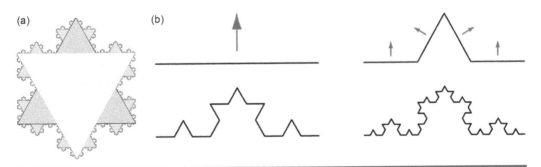

Figure 3.6 The Koch Snowflake. (Courtesy Wikimedia Commons.)

mathematician Benoit Mandelbrot first published on these phenomena in his book *Fractals: Form, Chance, and Dimension*. Mathematically, a fractal is a self-similar entity, so if you zoom in on a section of it, the section will be perfectly identical to the whole.

Whether we are talking about the Fibonacci sequence or any other fractal pattern in nature, we are not seeing true fractals, but we see patterns that sufficiently mimic self-similarity by demonstrating it up to a finite point. Seemingly-fractal patterns can be expressed numerically the same way as hypothetical "true fractals." These pseudo-fractals are usually called "natural fractals."

The concept of fractals is vital to understanding ammonoids, but the spiral itself is not the only manifestation of fractal geometry on these fossils. Another feature of the ammonoid shell—ammonoid sutures—are another of many natural phenomena that can be described as natural fractals and represented fractally in math. This is because the septa, or the walls dividing their chambers, were folded. This is actually quite an important factor which distinguishes ammonoids from most nautiloid shells, as nautiloid septa are smooth. Where the edge of each septum meets the outer shell wall, a junction was formed which we call a suture (Figure 3.8). If you have an ammonoid shell with the outer wall eroded or polished away, the suture becomes visible (Figure 3.7), and the pattern formed at this junction appears fractal.

Primitive ammonoids such as goniatitids displayed relatively little folding, as they evolved from a simpler nautiloid ancestor, but over evolutionary time, the number and configuration of folds in ammonoid sutures became more and more elaborate and complex. Sutures also became more complex over the life cycle of individual animals.

Folds are given specific names based on the direction in which they pointed, and where they were located on a shell. Lobes pointed toward the shell's apex, or the original protoconch, and saddles pointed toward the aperture. The lobe that runs along the venter or keel of the shell is the ventral, or external lobe (E). The next lobe out is the lateral lobe (L). This lobe is usually rotated close to 90° around the shell from the lateral lobe, unsurprisingly, on the lateral flanks of the shell. Any lobes which follow L are the umbilical lobes (U), as they are tucked in toward the inside of the coil, near the umbilicus. There can be anywhere from 1 or 2 umbilical lobes to upwards of 15. Because there are usually multiple umbilical lobes, they are numbered. The first umbilical lobe (U_1) is the furthest from L, and the last one [(U_3), if there are three umbilical lobes, or U_{13} if it is the 13th] is closest to L.

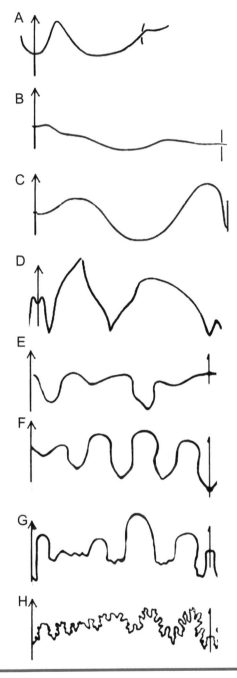

Figure 3.7 **(a) Nautilitic suture. (Redrawn from USGS.) (b) Bactritid suture. (Redrawn from Gordon 1966.) (c) Agoniatitic suture. (Redrawn from Miller et al., 1957.) (d) Goniatitic suture. (Redrawn from Korn and Ebbighausen, 2008.) (e) Clymeniid suture. (Redrawn from Treatise Vol. L.) (f) Prolecanitid suture. (Redrawn from Treatise Vol. L.) (g) Ceratitic suture. (Redrawn from Treatise Vol. L.) (h) Ammonitic suture. (Redrawn from Shevyrev, 1963.) The arrow on the left shows the location of the venter (outer edge of shell), and points toward the apex of the shell. It lies in the middle of the ventral lobe. Each lobe (pointed toward apex) or saddle (pointed toward aperture) is then numbered from the venter. The line on the right marks the position of the umbilicus in the suture.**

While ammonite paleontologists normally rely on lobes, and not saddles, to describe locations on a shell, the saddles also have numbers. The saddle numbering runs in the opposite direction of the numbering of the umbilical lobes, such that the first saddle (S_1) is between E and L. S_2 is adjacent to the last umbilical lobe, and the subscript numbers increase as the saddles get closer to U_1.

Lineages of ammonoids developed their own distinct characters in the geometry of these elements. The lytoceratoids, a superfamily from the Jurassic and Cretaceous, have boomerang-shaped saddles and lobes, and most are folded several times over. The phylloceratids ("leaf-horns" in Greek) of the Early Triassic tend to follow a pattern of tridentine saddles and oak-leaf-shaped lobes. Phylloceratoids were the first ammonites, so the round saddle and jagged lobe was retained from the ceratitids, their direct ancestors.

Evolutionary biologists such as Stephen Jay Gould have closely studied the connection between increased complexity and increased variation within a given lineage. As he pointed out, all organisms arise from a state of minimal complexity but become more complex over multiple generations. As they become more complex, variations on the same theme emerge.

And so it was in the case of ammonoids. Ammonoids with low septal complexity, such as goniatitids, show little variance in the geometry of their sutures. Specimens from the same family, or sometimes different families, may have virtually indistinguishable septal geometry. This changed by the Permian, when the frilly sutured ceratitids appeared. Suddenly, almost no geometric features were in common at the family level, and the genus became the new broadest level at which distinct septal features could be considered endemic or unique to a group of ammonoids.

As complexity of the suture pattern increased in Mesozoic, so did the amount of variation (Figure 3.8). Two- and three-tined saddles and lobes gave way to elaborate patterns of intricate folds. As the complexity is increased, so is the amount of variation–lopsided saddles and lobes, and sutures growing right on top of one another became common. Handedness—one side of an ammonoids' body having more complex septal geometry than the other—became common as early as the Triassic.

There has long been intense speculation about the specific function of these complicated sutures beyond the basic use of chamber walls seen in living nautiluses today, but certain things are constant. The more complex the suture pattern, the thinner the wall between the shell becomes.

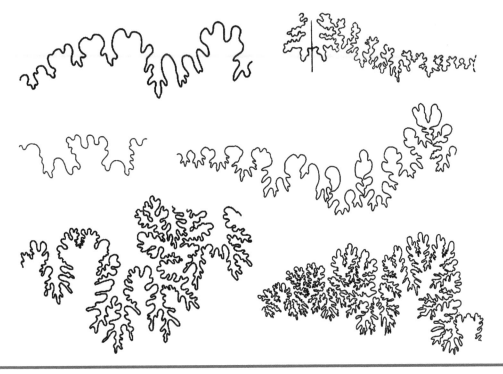

Figure 3.8   Ammonite sutures from the Jurassic and Cretaceous. (Courtesy USGS.)

In the past, the increasingly complex sutures of progressively later and more derived ammonoids has been interpreted to be a byproduct of the rapid lifespans of many ammonoids. It turns out that depositing new shell is an energetically costly process for mollusks. As nautiluses live about 20 years, and most ammonoids lived about 4 years, how else would ammonoids have the time and energy to deposit a strong shell? Just a couple of years could surely not be enough time for the thick, smooth deposition of shell seen in nautiluses. Lytoceratid ammonites in particular, which often have highly complex sutures with Y-shaped saddles and lobes, were exemplified as having increased their tensile strength without adding thickness primarily for the behavior of vertical migrations. Nautiluses undertake vertical migrations daily, but they are protected from sudden changes in water pressure by relatively thick shells. Lytoceratitic shells, in contrast, were much thinner.

There may not be a single environmental driver of septal complexity. Some pelagic ammonites, like placenticeratids, had relatively undivided sutures. In diplomoceratids, paperclip-shaped ammonites with sutures like those of the lytoceratids, highly complex sutures are associated with floating species, and simpler sutures are associated with resting on the seafloor (Figure 3.9).

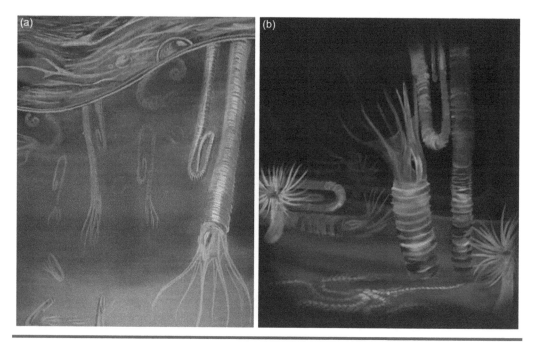

Figure 3.9 **Benthic and semi-pelagic diplomoceratid ammonoids.**

Fractal Geometric Methods

There are several methods for getting the fractal dimension of an ammonoid suture. One of the oldest, as well as the most common, comes from the measurement of England's rocky coastline (Figure 3.10). The craggy, uneven shoreline of England makes it difficult to determine the shoreline's true length, so it is measured as its jagged pattern moves inland and out to sea. To do this, a rule size is chosen. Let's say we have a meter stick, or for our purposes, 100 cm.

To measure an unevenly eroded beach in England (which is fitting, as jagged English beaches are often filled with ammonites), we take our meter stick and lay it down at Point A. We then pick it up and place it back down, end-to-end from where it was originally placed. We do this all the way up the beach, changing our angle if the shoreline's perimeter dips inland or outward toward the sea.

This method is called the Coastline Paradox because a straight-line distance is not achieved. Instead, we have a distance that reflects the many small curves, nooks, and crannies of the eroded shoreline. Let's say our connect-the-meter-stick distance from Point A to Point B is 280 m. But given that it's not a straight-line distance, what if we just want to know how complex the shoreline's jagged edge is?

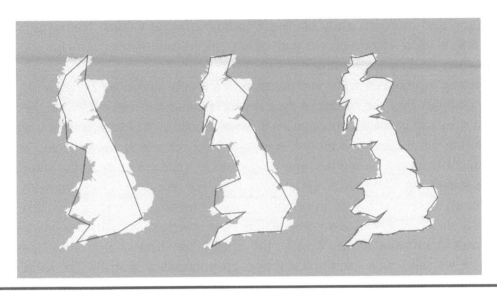

Figure 3.10   The fractal Coastline Paradox of the full perimeter of England. (Modified from Acadac.)

We can take the fractal dimension of the shoreline using the data we have just collected in a simple logarithmic equation. We placed the meter stick down 280 times, and there are 100 cm in a meter. (We can't use meters, because log(1)=0, and our answer will be undefined.) So we divide the log of 280 by the log of 100 to solve for our fractal dimension, or *Df.*

$$Df = \log(280)/\log(100) = 1.224$$

Say we measured the same shoreline distance from Point A to Point B using a 50 cm ruler instead of a meter stick. As the ruler is smaller, we will count it more times along the same path. When we enter the new data into our equation, both the high number of times we used our ruler, and the low centimeter count will dramatically change the fractal dimension:

$$Df = \log(690)/\log(50) = 1.671$$

The number is not always this much higher when we use a smaller ruler, but it is more accurate. Looking at the Coastline Paradox (Figure 3.10) we can see that the smaller ruler size offers greater resolution than the larger one; it more closely hugs the true perimeter, while the larger ruler size smooths out and generalizes the shape, leaving more fine details out.

The same thing can be done, on a much smaller scale, with ammonite sutures, and the same rules apply: a smaller ruler size offers greater accuracy, and more ins and outs result in larger fractal values. The same equation is used. However, we generally work in millimeters, not centimeters, when dealing with ammonites.

Because ammonites grew, and therefore, each new suture got bigger and more complex over the course of their lives, choosing our ruler based on the actual ruler can create inconsistencies if we are looking at ontogeny. If you were measuring sutures within a growth sequence, then, choosing a ruler length of 10mm, for example, would become progressively smaller relative to the suture it was measuring, and therefore, more accurate. For researchers working with sutures on a computer, scale information may be missing altogether. For these reasons, since the 1990s, many researchers have switched from an absolute measurement to a relational one. We take the straight-line distance from E to U_1—a distance we call *Lmax*—and divide it by a given number, often 10. The reciprocal of that number, in this case, 1/10 of *Lmax*, becomes our ruler.

For a suture from *Eogaudryceras numidum* (Figure 3.13), the ruler length of 1/10 Lmax fit the curve of the suture 47 times. To determine our fractal dimension, we divide log(47) by log(10) (Figure 3.11):

$$Df = \log(47)/\log(10) = 1.342$$

Given that the geometry of ammonoid sutures is incredibly varied, what features determine how complex an ammonite suture will be? We know that as complexity increases, septal thickness decreases. The number of saddles and lobes is a considerable factor; as new ones are added, there is less space for division inside individual saddles and lobes. Ammonites like *Placenticeras*, which usually have a high number of umbilical lobes have smaller lobes and saddles overall, have smaller saddles and lobes, whereas ones like *Lytoceras*, which usually have three umbilical lobes or less, have much larger saddles and lobes overall. Statistically, larger (and less) saddles and lobes are more heavily divided. A near-perfect converse variation exists such that the larger the lobes (and less umbilical lobes), the further the septal subdivisions go.

Likewise, the mechanism by which these patterns even formed has also eluded paleontologists. For decades, the question of what made ammonites produce these patterns (given that no other shelled cephalopod does anything like this) has puzzled us and become a subject of intense debate. There were two prevailing (read: warring) hypotheses for most of the twentieth century. Elements of either possibility continue to resurface periodically, with some more recent researchers combining pieces of both.

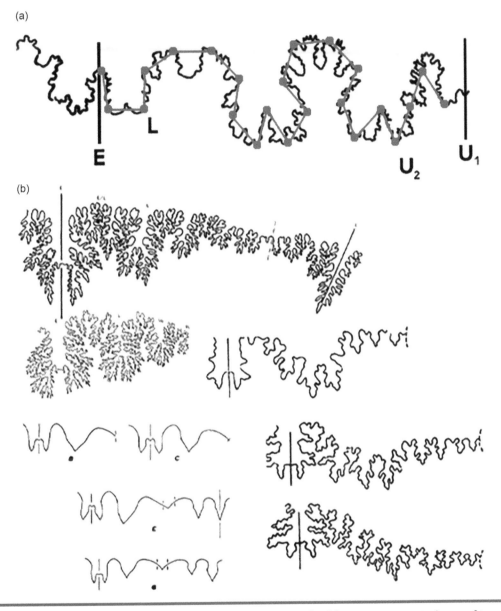

Figure 3.11   (a) Suture of *Baculites grandis* measured with 1/10 Lmax 22 times. (b) Sutures of various ammonoids. (Courtesy USGS.)

The first of these is the tie-point theory. It posits that while the dissolved calcium carbonate being deposited and solidifying behind the creature's soft mantle, sinuous strands of tissue ran the perimeter of the new chamber wall and pulled the newly forming septum into place. This model, called the Tie-Point Model, was championed by the legendary German paleontologist Dolf Seilacher, and to a lesser extent by his contemporary, Gerd

Figure 3.12   The Viscous Fingering Model, or Saffman-Taylor Instability TiO$_2$ gel. (Courtesy Claire Trease, Kingston University.)

Westermann. In exceptionally preserved specimens, a series of soft-tissue strands are found behind the mantle, and they begin to separate as one looks closer and closer to the last septum (Figure 2.1d). However, no one has been able to determine exactly what these threadlike structures are. Not surprisingly, the model is not without its problems. The glaringly obvious issue is that in even exceptionally preserved (and prepared) specimens, attachment points for the strands have not been found.

For most of the 20th century, the main competition for tie points was the Viscous Fingering Model (Figure 3.12). This hypothesis hinges on the physics phenomenon known as the Saffman-Taylor Instability, in which two fluids of different viscosities are brought together but are unable to mix due to their consistencies. Instead of combining, the two fluids interface in a push-pull configuration, between one another, and forming dendritic patterns.

Proponents of the Viscous Fingering Model suggest that the partially dissolved calcium carbonate acts with the cameral fluid, seawater, and perhaps the internal body chemistry of the animal itself in this manner, resulting in fractal patterns. Similar phenomena are seen in batteries where fluids of different ionic concentrations form dendrites rather than combining. Supporters of the model point to ontogenetic and pathological data—the unsteady increase in sutural complexity over ontogeny could reasonably support the formation of sutures by a chemical reaction between liquids. As with the Tie-Point Model, however, Viscous Fingering is lacking in definitive evidence, and by itself, it fails to address several questions. The most glaring is how can such a simple inorganic process account for the incredible disparity of diverse suture patterns? Surely there is a genetic component to the production of these patterns, as distinct patterns belong to specific lineages. Calcium carbonate and seawater could not, by themselves, consistently produce these complex familial distinctions without a genetic component.

Years after the appearance of both models, a third model, the Cerata-Septa Model, came onto the scene. This model looked to living mollusks to attempt to parse out the elusive formation of ammonoid sutures. Shinya Inoue and Shigeru Kondo examined the mantle structures of modern sea slugs belonging to the aptly named superfamily Dendronotoidea. Over the life of the sea slugs, the mantle structure increases in its fractal complexity through branching of the soft tissue into protrusions called cerata (hence the name of the superfamily). The team then used CT imagery to compare the branching patterns of the dendronotoid sea slugs to the septa of *Damesites*, an ammonite from the Cretaceous. They found enough similarities to suggest that septal complexity is driven by branching that occurs in some mollusk's rear mantle, potentially in the mantle of ammonoids. This model supported neither the Tie-Point nor the Viscous Fingering Models, and instead corroborated lesser known, poorly examined, and somewhat forgotten models from decades ago that suggested an increasingly frilly mantle "stamping" itself on the new septum as it crystallized. It addresses a key limitation of both Tie Points and Viscous Fingering: because families of living mollusks that exhibit branching have distinct patterns, the question of how a given suture pattern was passed down over generations of ammonoids would be in line with every other branching mollusk (Figure 3.13). Furthermore, hollow, straw-like holes at the edges of the septum act as strong evidence for a branched mantle. However, not all examples of

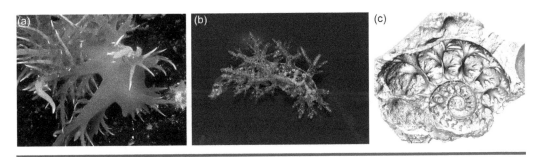

Figure 3.13   **Branching sea slugs from the superfamily Dendronotoidea. Note the similarity between these branching patterns and the CT scan of the ammonite *Gaudryceras*. (a) *Dendronotus iris*, (b) *Dendronotus frondosus*, and (c) CT scan of shell interior from the ammonite *Gaudryceras*. (Courtesy Wikimedia Commons.)**

soft-tissue preservation show the branched pattern, and none show exceptionally high-order complexity. Is it possible that ammonoids changed their skin texture via papillae in their skin like modern cuttlefish?

Even Cerata-Septa has unanswered questions, although many current researchers have thrown their support behind it, due mainly to the fact that not enough soft tissue exists from ammonites to either confirm or deny it. Five years after Inoue and Kondo published the Cerata-Septa Model, Lesley Cherns and colleagues published a specimen of *Sigaloceras*, a Jurassic ammonite, which showed exceptional internal preservation of the soft tissue. The specimen is filled with translucent precipitated calcite, and backlighting it allows the viewer to clearly see its gills, digestive organs, and potentially one or two muscles that held the animal in place in its body chamber. Despite unprecedented preservation of the mantle cavity, the ammonite did not appear to clearly favor any one mechanism of septal formation over the others, but it was not for a lack of possibilities. Viscous fingering would be impossible to prove with only one specimen, and modeling the interaction of fluids is necessary. Tie points could be argued, as the sinuous strands appear at the rear of the mantle, and they begin to separate in the final chamber after the last formed septum. However, very importantly, there is not a strand interacting directly with the last septum: the specimen does not appear consistent with the Tie-Point Model. Finally, the fleshy mantle itself is missing, so we cannot see how the mantle interacted with the last septum. The sinuous strands are positioned between where the mantle belongs and the last chamber.

The underlying reason for suture complexity has also been considered equally murky. While we no longer believe that suture geometry had a direct relationship to the habitat or type of mobility of specific ammonoids, the differences in their septal geometry undoubtedly indicate biological differences.

The exact underlying mechanism may never be fully understood, but the fractal geometry of ammonites may yet tell us much about the way these animals lived. Gerd Westermann believed that the complex suture geometry of *Lytoceras* meant it was a vertical migrator, and that its complex sutures aided in structural support to the shell without adding thickness. Supporters of this idea argued that because ammonites probably lived much shorter lives than nautiluses, increasing complexity could increase the shell's tensile strength—its resistance to being crushed—without the need for time-consuming and energetically costly deposition of thick calcium carbonate layers. This argument persisted for the entire second half of the twentieth century.

While it is true that depositing thick layers of calcium carbonate shell is incredibly draining on the energy of mollusks, and that with their short lives, ammonites like *Lytoceras* would have had had less time to deposit the same type of shell as a *Nautilus*, other functional explanations are now considered more likely.

Further Reading

Archimedes. On Spirals. https://www.aproged.pt/biblioteca/worksofarchimede.pdf.

Burnaby, T. (1966). Allometric Growth of Ammonoid Shells: A Generalization of the Logarithmic Spiral. *Nature*, 209, 904–906. https://doi.org/10.1038/209904b0.

García-Ruiz, J.M., Checa, A. (1993). A Model for the Morphogenesis of Ammonoid Septal Sutures. *Geobios*, 26, 157–162. doi:10.1016/s0016-6995(06)80369-4.

García-Ruiz, J., Checa, A., Rivas, P. (1990). On the Origin of Ammonite Sutures. *Paleobiology*, 16(3), 349–354. doi:10.1017/S0094837300010046.

Lutz, T., Boyajian, G. (1995). Fractal Geometry of Ammonoid Sutures. *Paleobiology*, 21(3), 329–342. doi: 10.1017/S0094837300013336.

Mandelbrot, B. (1982). *The Fractal Geometry of Nature*. W. H. Freeman and Co., New York.

Marriott, K. (2022). Lateral Data Normalizations for Partial Goniatitic and Ceratitic Sutures. *Bulletin of the New Mexico Museum of Natural History*.

Meisner, G.B., Araujo, R. (2018). *The Golden Ratio: The Divine Beauty of Mathematics*. Race Point Publishing, New York.

Olóriz, F., Palmqvist, P. (1995). Sutural Complexity and Bathymetry in Ammonites: Fact or Artifact? *Lethaia*, 28(2), 167–170. doi: 10.1111/j.1502–3931.1995.tb01608.x.

Thomson, D.W. (1917). *On Growth and Form*. Cambridge University Press.

Chapter 4

The First Cephalopod: A Snail Who Dared to Fly

The state of the world at the time life started was very different from the world we know today. In many ways, the early biosphere, evidenced by our rock record, was so inhospitable that it is hard to imagine anything could have survived in it. The oldest known fossils are of single-celled microorganisms dated to about 3.5 billion years. The atmosphere was made mostly of nitrogen and carbon dioxide, but no free oxygen was present. An oppressively bright sun blazed down on Earth with no ozone layer to penetrate on its way into the young planet's atmosphere. Ultraviolet rays bathed the land surface with such an intensity that any modern land creature would die quickly. All life was in the sea, which screened out much of the deadly ultraviolet. In turn, the land was largely a desolate mix of dull brown and gray, with only the occasional sparse patch of primitive blue-green bacteria. Fine, dry clasts of inorganic sediments—sand, clay, and silt—blew lifelessly in the wind. In many places, the foundations of what would someday be Earth's continents are so fresh that these masses, the cratons, are still largely exposed—an infernal landscape that is far from stable, let alone nurturing.

It took almost two billion years for multicellular organisms to evolve from cells floating haplessly in the oceans of this seemingly desolate planet, to the stage where some consumed others (heterotrophy) evolved. All of the first organisms were chemoautotrophic, deriving their energy from inorganic compounds floating around them in the marine environment. Eventually, some blue-green bacteria developed the ability to take sunlight and carbon dioxide and turn it into organic matter. The exact origins of complex,

DOI: 10.1201/9781003288299-4

multicellular life are lost in the mists of time. Multicellular life is thought to have three possible, not mutually exclusive causes, all of which have living analogues.

Colonial Organisms

The first possible impetus for multicellular life were colonies of multicellular organisms, which occurred occasionally in the late Precambrian. These organisms evolved specialized roles within the colony. A modern example of this would be the Portuguese Man-O-War, a colonial organism which resembles a sea jelly. Part of a group of animals called siphonophores (Figure 4.1), each constituent animal in a Man-O-War is only one cell, and each only

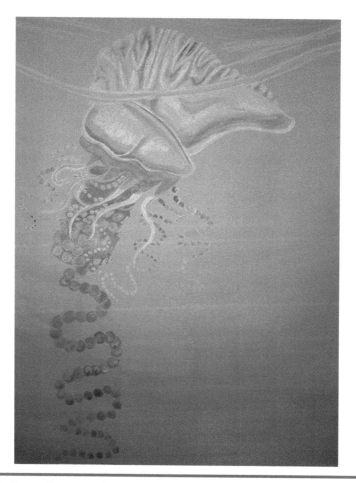

Figure 4.1 **Siphonophores with specialized individuals in a Portuguese Man-O-War colony.**

performs one task. Tasks include movement, catching food, producing food, converting food into energy, and buoyancy control.

Another example of this is sponges. Each sponge is built of a "skeleton" of needle-like structures known as spicules, woven together by the individual cells of the sponge. But each individual cell is completely independent of the others in all other respects except for secreting the spicules to build their common skeleton. Each sponge cell can do everything it needs to live independently: respiration, feeding, reproduction, and flailing their whip-like flagella to drive the water currents past them, so they can trap tiny food particles and oxygen. These individual cells of the sponge animal are so independent that if you force them through a fine sieve and break them up into separate cells, they can reassemble to form a new sponge after a number of hours.

Syncytial Cells

Of the three models for an origin to multicellular life, this is the least supported by biologists. Some organisms have cells with more than one nucleus—a phenomenon known as syncytium (Figure 4.2). This is a fast, low-energy way to duplicate living material. Some living slime molds are syncytial. However, the reason this possible origin story is not well supported is because there is no known example of the specific *type* of syncytium that would need to exist for multi-nuclei to result in multicellular life.

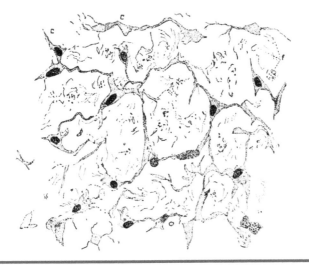

Figure 4.2 **Syncytium in a slime mold. (Courtesy Wikimedia Commons.)**

Each nucleus would need to develop its own cell membrane and ultimately, its own specific function within the cytoplasm mass. No living or fossil species has been shown to do this.

Symbiosis: The Powerhouse of the Cell

A partnership between cells, known as a symbiosis (Figure 4.3), is about the most well-supported possible origin for multicellular life. When one of the symbiont cells lives inside the other symbiont, we call this endo-symbiosis. A cell eventually enveloped another cell in order to inherit its energy for the first time. The smaller cell was destroyed in the process, and the first instance of an organism eating another led to a specialization of single-celled organisms which preyed on their neighbors. Eventually, this new evolutionary pressure was met with prey organisms who could withstand being eaten. After generation upon generation of prey organisms who could survive being eaten, a symbiosis emerged: the prey organism eventually lived inside the predator organism, producing energy for it in exchange for shelter. Several known examples of this exist. Zooxanthellae are single-celled photosynthetic algae called dinoflagellates which live inside

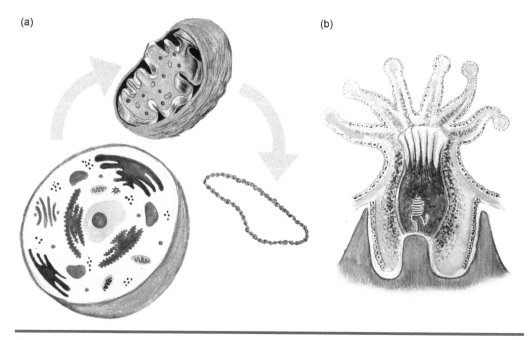

(a) (b)

Figure 4.3 **(a) Endosymbiosis in a coral polyp and (b) mitochondria, with mtDNA and a separate membrane, inside an animal cell. (Courtesy Wikimedia Commons.)**

transparent coral polyps. Usually, dinoflagellates possess flagella, whip-like appendages that allow them to swim through the water. However, zooxanthellae inside of a coral polyp have lost their flagella, adapting (sometimes in just a generation or two) for life inside a coral polyp. The coral provides protection to the zooxanthellae, and the zooxanthellae consumes carbon dioxide and provides oxygen for the coral. The zooxanthellae deposit the polyp's share of glucose directly into the fleshy lining of the host's stomach cavity.

In 1967, the biologist Lynn Margulis proposed that endosymbiosis might explain the origin of multicellular organisms. Instead of the difficult process of evolving organelles out of nothing, Margulis argued that each of the organelles found in the eukaryotic cell were once free-living prokaryotes that had come to live symbiotically within another cell, and eventually become part of it. Chloroplasts apparently started out as blue-green bacteria, which are photosynthetic, even though they are prokaryotes without organelles. Purple non-sulfur bacteria have the same overall structure and function as mitochondria, and apparently that's where these organelles came from. The flagellum of many cells has the identical $9+2$ fiber structure (nine sets of microtubule doublets surrounding a pair of single microtubules in the center) as the prokaryotes known as spirochaetes, which also cause the disease syphilis. As each of these smaller prokaryotes came to live within a larger cell, they sublimated their functions to that of their host, so that the cyanobacteria became chloroplasts that are now production centers for photosynthesis. The purple non-sulfur bacteria became mitochondria and performed the role of the energy converter for the cell.

In addition to the detailed structural similarities of these prokaryotes and the organelles of eukaryotes, Margulis pointed to many other compelling lines of evidence. Usually, organelles are not floating within the eukaryotic cell membrane but separated from the rest of the cell by their own membranes, strongly suggesting that they are foreign bodies that have been partially incorporated within a larger cell. Mitochondria and chloroplasts make proteins with their own set of biochemical pathways, which are different from those used by the rest of the cell. Chloroplasts and mitochondria are also susceptible to antibiotics like streptomycin and tetracycline, which are good at killing bacteria and other prokaryotes, but the antibiotics have no effect on the rest of the cell. Even more surprising, mitochondria and chloroplasts can multiply only by dividing into daughter cells like prokaryotes, and thus have their own independent reproductive mechanisms; they are not made by the cytoplasm of the cell. If a cell loses its mitochondria or chloroplasts, it cannot make more.

When Margulis' startling ideas were first proposed over 60 years ago, they were met with much resistance. But as biologists began to see more and more examples of symbiosis in nature, the notion became more plausible. We humans have millions of endosymbionts (mostly bacteria) on our skin and inside us. Our intestines are full of the bacterium *Escherichia coli* (*E. coli* for short), familiar from Petri dishes and news alerts about sewage spills or contaminated kitchens. These bacteria actually do most of our digestion for us, breaking down food into nutrients in exchange for a home in our guts. Most of our fecal matter is actually made of the dead bacterial tissues after digestion, plus indigestible fiber and other material that we cannot metabolize. There are many other examples of endosymbiosis in nature. Termites, sea turtles, cattle, deer, goats and many other organisms have specialized gut bacteria that help break down indigestible cellulose, so these animals can eat plant matter efficiently. Tropical corals, large foraminifera, and giant clams all house symbiotic algae in their tissues, which produce oxygen, remove carbon dioxide, and help secrete the minerals for their large shells.

The strongest evidence came when people started studying the organelles more closely and found that not only did they have the right structure to have once been independent prokaryotic cells, but they also have their own genetic code! Mitochondria and chloroplasts both have their own DNA, which has a different sequence than the DNA found in the cell nucleus. In fact, the mitochondrial DNA is different enough and evolves at a different rate from nuclear DNA, so it can be used to solve problems of evolution that the nuclear DNA cannot. This would make no sense if the eukaryotic cell had tried to generate the organelles from scratch. They would not have their own genetic code if that were true.

The final clincher in the endosymbiosis hypothesis is that there are many living endosymbiotic cells which show that this process occurs even today. The simpler eukaryotes, such as the freshwater amoebas *Pelomyxa* and the *Giardia* (famous for causing dysentery from contaminated water), lack mitochondria but contain symbiotic bacteria that perform the same respiratory function. In the laboratory, scientists have observed amoebae that have incorporated certain bacteria in their tissues as endosymbionts. The parabasalids, which live in the guts of termites, use spirochetes for a motility organ instead of a flagellum. Thus, from the wild speculation of 1967, Margulis' idea is now accepted as the best possible explanation of the origin of eukaryotes and organelles.

The Cambrian "Explosion"

The emergence of the basic body plans for all invertebrate life forms occurred about 540 million years ago during the Cambrian. The Cambrian Period is not the first biological revolution in Earth's history, but it is the definitive beginning of the Phanerozoic Eon, a period of time characterized by the presence of complex organisms and ecosystems that continues to this day. Soft-bodied multicellular organisms began to diversify rapidly. In just tens of millions of years (a blip in geological time), the algae-dominated oceans ushered in unprecedented biodiversity. Demersal and pelagic niches emerged. With these new niches, soft-bodied animals evolved hard parts. Some organisms evolved the ability to selectively crystallize calcium carbonate, resulting in shells.

For the pioneering European geologists in the 1800s, very little was known of the early fossil record. The lack of suitable fossil-bearing sedimentary rocks below the oldest fossiliferous strata (then called "Silurian," but now recognized as Cambrian) was a puzzle to them. To the early geologists, the apparent absence of Precambrian fossils, and then the apparently rapid of diverse trilobites in the Lower Cambrian strata, seemed abrupt, and so the mistaken term "Cambrian explosion" was born. Of course, Precambrian strata are *not* unfossiliferous—they are full of microfossils (in rocks like cherts that preserve them well), but lack any megascopic fossils except stromatolites, whose biological origin was not confirmed until just 70 years ago. Thus, the absence of Precambrian fossils was an illusion; nineteenth-century geologists were mistaken in their expectation to find megafossils in strata that had only abundant microfossils.

But elsewhere in the world, there are strata that preserve the latest Proterozoic marine conditions well, and they *do* produce megascopic fossils. The first to be described were the deposits of the Ediacara Hills in the Flinders Ranges of southeast Australia. These date to 600 Ma, and they are known as the Ediacaran fauna (pronounced "Ee-dee-AKK-ara"). This period of time from 600 Ma to the beginning of the Cambrian 545 Ma is known as the Ediacaran Period of the Proterozoic Era. The Ediacara fauna is now known from a wide variety of localities around the world, including many spectacular localities in China, Russia, Siberia, Namibia, England, Scandinavia, the Yukon, and Newfoundland. Most of these fossils are the impressions of soft-bodied organisms without shells, so they had no hard parts that make up the bulk of the later fossil record. Instead, these impressions have reminded some paleontologists of the impressions made by jellyfish, worms, and soft corals and other simple non-skeletonized organisms.

Although the Ediacaran fauna clearly represents fossils of multicellular organisms (some reach almost a meter in length), paleontologists have a wide spectrum of opinions about what made these impressions. The older, more conventional interpretation is that they are related to groups we know today: sea jellies, sea pens, and worms of various sorts. Some do look a bit like a sea jelly, but if so, they have symmetry unlike any living form. Others vaguely resemble some of the known marine worms, although their symmetry and segmentation does not match any groups of worms alive in the ocean today. Nor do the "worms" have evidence of eyes, mouth, anus, locomotory appendages, or even a digestive tract. For this reason, other paleontologists have suggested that Ediacaran fossils were made by organisms unlike any that are alive today. Whatever the biological affinities of the Ediacara fauna, it is very clear that they are some kind of multicellular organisms, whether animals, plants, fungi, or some early experimental organisms not related to any living group.

So, the old, outmoded concept of the "Cambrian Explosion" of abundant trilobites following absolutely nothing is now completely out of date. We now have three billion years of microfossils, then at 600 Ma the appearance of the first megascopic soft-bodied multicellular organisms, the Ediacara fauna. Even though some of them were large, they still didn't have modern body plans, nor did they have any hard parts. The next evolutionary step was to develop some sort of shelly framework. This also means that these organisms would leave behind the first hard-part fossils. As you would expect, the first shelled organisms acquired a shell in small steps, forming tiny (less than 1 mm) hard parts nicknamed "little shellies" or "small shellies." Many of these "little shellies" appear to be primitive cap-shaped mollusks or even clam-like shells; others appear to be bits of "armor" that made up a "chain-link" covering on some animals; still others (like the tiny spiky "maces" known as *Chancelloria*) are spicules of creatures that may be related to sponges. The affinities of most of the "small shellies" are still largely unclear. They appeared in the first series of the Cambrian Period (Terranuevian Stage, beginning at 545 Ma). The megascopic skeletons of trilobites finally appeared in the second Cambrian Stage, the Atdabanian, about 520 Ma.

Thus, the term "Cambrian explosion" is obsolete and inaccurate. It was proposed when we knew almost nothing about the many different evolutionary steps (from single-celled to multicellular soft-bodied Ediacarans to "little shellies") in the origin of large skeletonized fossils like trilobites. If you look at it from the perspective of 520 million years later, then the 80 m.y. between 600 and 520 Ma, or the 25 million years of the first stage of the Cambrian, might look a little bit like an "explosion." But 80 million years of time from the Ediacaran to the Middle Cambrian (longer than the entire Cenozoic), or even 25 million years of the Early Cambrian, is hardly an "explosion" by any objective standard. Geologists are gradually abandoning the misleading and obsolete term "Cambrian explosion" when it was clearly a "Cambrian slow fuse."

Mollusks Arise

The third stage of the Early Cambrian is most commonly associated with the sudden appearance of thousands of early arthropod species, especially trilobites. The period was equally significant for mollusks. The first mollusk-like organisms appeared in the Precambrian. *Kimberella* (Figure 4.4) was a common constituent of the Ediacara fauna, which appeared about 555 million years ago. Whether or not *Kimberella* were true mollusks is still speculated, but they are generally accepted to have been Bilateria—a wide clade of nearly all animals, both invertebrate and vertebrate, which are characterized by bilateral symmetry.

The first true mollusks appeared about 545 Ma, shortly before the beginning of the Cambrian Period. Helcionellids (Figure 4.5) have many traits in

Figure 4.4 *Kimberella,* a possible first mollusk. (Courtesy Masahiro Miyasaka.)

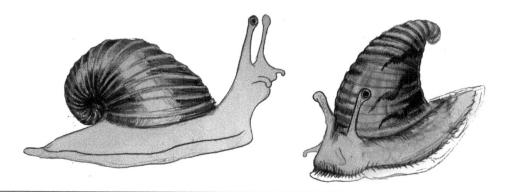

Figure 4.5 **Evolution of slight coiling in helcionellids.**

common with primitive snails. Coiling is a specialized trait in many mollusk shells of all kinds, but the ancestors to gastropods (or univalves) were monoplacophorans, whose shells exhibit almost perfect flatness, like the domed shape of a limpet. Helcionellid shells are conical and exhibit little coiling along an axis. Most helcionellids were only a few millimeters in diameter and may have been similar in anatomy and behavior to limpets, but there is debate over whether or not they are true gastropods.

Cephalopods Emerge

By the Middle Cambrian, crawling gastropods, echinoderms, and trilobites populated the benthic realm of warm, shallow seas. These fossils show us an ocean floor littered with sea pens and crinoids, an anchored, distant relative of sea stars. It was churned by the tracks of trilobites, early relatives of crustaceans and arachnids, as bountiful in their own world as city cockroaches in summer. There were no fish. Worms burrowed up and down through the sediment, and sedentary animals undulated slowly in the ocean current. With increased competition on the seafloor, animals began to expand their territory into the water column above. A few creatures adapted to glide above and harvest the seafloor scuttlers: the sea jellies and some arthropods did it first. Swimming arthropods, including the early super-predator *Anomalocaris,* emerged, as well as free-swimming trilobites, and bizarre, unrepeated organisms like the five-eyed opabiniids.

Then a small, strange creature in a conical shell took to the open sea.

Gastropods became abundant during the Cambrian, but during their unchecked biodiversification, a subset of them developed a novel way to float up into the water column. These intrepid, glorified snails sealed off the back end of their shell, save only for a thin tube of tissue with hypertonic blood. With the salty blood absorbing all the moisture in the sealed chamber, it filled with gas, and the animal could float upwards. With this adaptation, the siphuncle, the first true cephalopod had learned to control its movement along a third axis. They began to detach from their rocky substrates, and by the end of the Cambrian, a new subclass of mollusks had evolved from the gastropods, one whose primary mode of life was floating and grasping. The fossil known as *Plectronoceras* (Figure 4.6) is considered the oldest known fossil cephalopod. Its soft tissue probably resembled something between that of a nautilus and a small, pointy marine snail. Compared to later cephalopods, it probably had limited control of its movement in the water column: *Plectronoceras* shells contained just a few, very short chambers. Their phragmocone was still very small relative to the body chamber. With the soft-tissue-filled living chamber taking up half the shell, the floatation adaptation, though revolutionary, was limited.

Figure 4.6 **Plectronoceras.**

Despite its limitations, *Plectronoceras* was the first true nautiloid. A siphuncle threaded the center of each of its septal walls. From *Plectronoceras*, the cephalopods diversified and were able to join the soaring apex predators of their time in the water column. The nautiloids added new chambers, granting them further control over movement. These early nautiloids had straight shells, orthocones, or slightly curved shells, cyrtocones, that may have held various orientations and behaviors for various niches. More than likely, most of them held themselves horizontally and likely fed on trilobites and other arthropods, scaphopods, carrion, and even smaller nautiloids. It is unlikely that they were picky—cephalopods have likely always been both carnivores and scavengers—but in the Late Cambrian seas, these primitive cephalopods certainly had options.

Will the Real First Cephalopod Please Stand up?

With the 1933 discovery of the plectronocerids, pinning down the first cephalopod originally seemed simple. The image of all cephalopods descending from externally shelled nautiloids makes sense. However, piecing together evolutionary history requires the perpetual openness of researchers to reassessment when evidence emerges for competing narratives. The full story of how cephalopods came to be from gastropods is, therefore, still under construction.

In 1976, the Burgess Shale yielded a new series of bizarre creatures, the likes of which the world had *sort of* seen. *Nectocaris pteryx* (Figures 4.7 and 4.8) sat undescribed in the collection of Charles Doolittle Walcott for over 50 years. Though its name means "swimming shrimp," the animal was unlike the other Burgess fauna. It had no segments, so it was not an arthropod. It had tentacles, so it was not a worm.

Additional species of nectocarids emerged from the Chengjiang and Burgess Shale fossil localities. By 2010, 91 specimens of the soft-bodied, flattened *Nectocaris pteryx* were more than sufficient to enable paleontologists Martin Smith and Jean-Bernard Caron to reconstruct the creature. In their studies of nectocarid morphology, another clue emerged: like modern octopodes and squid, *Nectocaris* appeared to have an external siphon: the characteristic cephalopod hyponome.

The discovery of a possible siphon in *Nectocaris* had the potential to upend our understanding of early cephalopod evolution. If the nectocarids were indeed a stem group of Cephalopoda, they would push the onset of cephalopod evolution 30 million years older. Moreover, if nectocarids were cephalopods (whether stem cephalopods, or proper "crown group" ones) it would mean that jet propulsion evolved independently of the external shell.

Figure 4.7 **(a)** *Nectocaris pteryx* **fossil. Note the siphon-like protrusion to the left of its head.**

The shell and apparent jet propulsion, therefore, would eventually need to come together in a siphuncle and chambers (Figure 4.9).

Most recently, the possibility of the nectocarids as stem cephalopods has been dismissed by some paleontologists. A few people still believe that these little creatures, which have anatomy like so many different groups, may truly belong with the cephalopods afterall or, for that matter, any other group with which science is familiar.

The Terror of Ordovician Seas

As the Paleozoic carried on, nautiloids went through several large pulses of shell reorganization: they began as straight, then went through several

Figure 4.8   *Nectocaris pteryx* swimming in the Middle Cambrian seas.

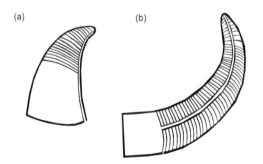

Figure 4.9   **(a) Cross section of shell of *Plectronoceras*. (b) Cross section of shell of *Cyrtoceras*.**

transitional phases of open gyrocones, and finally into the usually involute, closed spiral we all know today. They undid and redid this process several times, leaving discontinued evolutionary outliers at every pulse, and became incredibly diverse. The chambered nautiluses of today harken back to these primitive forms much more than a Mesozoic ammonite could ever hope to achieve. In terms of living creatures, the ammonite is probably best compared to the female argonaut octopus *Argonauta argo*, a coleoid floating freely in a paper-thin shell. Argonaut shells are not perfect analogues, however; the shell is not permanently attached to the animal. They also lack chambers, a siphuncle, and even a clear secretion point. The suture (structure separating two shell chambers) of the *Nautilus*, in contrast to ammonites, is a rounded C shape, and its concave side faces the aperture of the shell.

The most important change over the evolution of early nautiloid shells occurred in their chambers (Figure 4.9). *Plectronoceras* had short chambers, and so few of them that its body chamber had a greater volume than all its other chambers combined. Because of that, its swimming ability was much weaker than that of nautiluses we see today. It is believed that *Plectronoceras* could do little more than detach from its substrate and hover just inches above the seafloor for short periods of time. It probably had little control over its movements once in suspension. The cyrtocones elongated their chambers and their shells. As chambers became larger and over time, represented a greater volume of the shell, the siphuncle also gained power.

Orthocones are probably the best known of the early nautiloids. By the Ordovician, they evolved into a great diversity of different sizes and shapes. For a long time, it was thought that because of the way in which they secreted calcite in the phragmocone as a counter-weight to balance the buoyancy of the gas-filled chambers, that all orthocones were oriented horizontally in the water column. Some of them deposited calcite in the spaces in each chamber (endocameral deposits). Others had lots of extra calcite secreted around the siphuncle (endosiphuncular deposits). Others had chambers separated by septa shaped like cones, and these endocones were thick walled and nested in one another like paper cups, providing ballast and counterweighting. Because of this counterweighting, their enormous straight shells floated behind them horizontally in the water. Straight-shelled cephalopods have evolved numerous times, separately, over the course of over 400 million years. However, Utah researchers Kathleen Ritterbush, David Peterman and Nicholas Hebdon discovered that many of these orthocones did not have adequate counterweighting to be oriented horizontally, which made them similar in orientation to these nautiloids their distant ammonite descendants in the Cretaceous genus *Baculites,* which had no counterweighting, so the back of their shells pointed upward when in the water column (Figure 4.10).

By the Late Cambrian and Early Ordovician, increased competition forced evolutionary response from every lineage of animals. The cephalopods were no different. Many types of straight-shelled nautiloids swarmed the seas of the Ordovician, with the array of sizes and adaptations to swimming discussed above. A group of nautiloids, called the endocerids, evolved a new defense against the onslaught of emerging predators: huge size. The type genus of the endocerids (Figure 4.11), *Endoceras,* evolved shells over 10 m (33 ft) in length as documented by nearly complete shells, and some

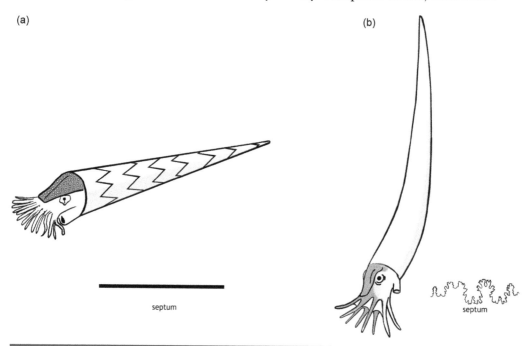

(a)

(b)

septum

septum

Figure 4.10 (a) *Orthocone* with substantial endosiphuncular deposits holding its orientation horizontally and (b) Actual orientation of many orthocones, as well as (Cretaceous) *Baculites.*

Figure 4.11 **The giant endocerid** *Cameroceras,* **merperson for scale.**

fragmentary shells suggest they reached almost twice that length. These massive shells made *Endoceras* longer than the largest giant squid living in the deep ocean today. *Endoceras* was untouchable by any other predator and reigned at the top of the food pyramid. These enormous cephalopods far exceeded the size of nearly any other marine organism of the time, which were typically a few centimeters in size. Even the largest trilobite of the Ordovician, *Isotelus rex*, was only 70 cm (28 inches) long, less than 10% of the size of *Endoceras*. There were small armored fish a few centimeters long, but they had no jaws, so they were no threat and instead were easy prey to medium-sized and large nautiloids. *Endoceras* was the terror of the Ordovician seas, easily grabbing nearly anything that swam near with its long tentacles, and pulling it in to its parrot-beaked mouth. A few lucky trilobites managed to escape this, as they show bite marks that might have been made by gigantic nautiloids.

But not every cephalopod adapted to the mounting environmental pressures with size. By the Silurian period, fish were starting to become bolder, more formidable forces. As this happened, the straight shells of cephalopods began to coil back up into tight planispirals. As fish developed jaws in the Silurian, more changes occurred in the cephalopods. These changes resulted in the very first ammonoids.

Further Reading

Davis, R.A., Meyer, D.L. (2009). *A Sea without Fish: Life in the Ordovician Seas of the Cincinnati Region*. Indiana University Press, Bloomington, IN.

Hildenbrand, A., Austermann, G., Fuchs, D. et al. (2021). A Potential Cephalopod from the Early Cambrian of Eastern Newfoundland, Canada. *Communications Biology*, 4, 388.

Kobayashi, T. (1987). The Ancestry of the Cephalopoda. *Proceedings of the Japan Academy, Series B*, 63(5), 135–138. https://doi.org/10.2183/pjab.63.135.

Miller, A.K. (1938). *Devonian Ammonoids of America*. The Geological Society of America.

Chapter 5

Colonizing the Wild Blue Beyond: The First Ammonoids in the Age of Fish

"Fossil Capitalism"

If you ever go to a rock shop or buy fossils online, among the most common fossils you see for sale are trilobites, ammonoids, and orthocone nautiloids from Devonian beds of the Tindouf Basin of Morocco in northwestern Africa. The region has some of the world's richest fossil sites, and since the late 1980s, it has become the major supplier of trilobites and other fossils on the international fossil market. Some 50,000 people earn their livelihoods by digging for the fossils, cleaning and polishing them (a skill called fossil preparation), and selling and trading them to collectors all over the world. The industry is worth some $40 million a year and is a major employer in some regions of Morocco. There are huge fossil trading areas between Erfoud and Alnif, in the Talfilalt region. In these places, the fossil trade is a leading creator of jobs. Specimens are shipped from the markets in Erfoud to larger Moroccan cities such as Casablanca, Marrakesh, and Rabat. The industry there is so enormous that it has been referred to as "fossil capitalism."

Most of the hard work of field collecting is done by local men who use picks, shovels, steel pry bars, chisels, and wedges to pry out slabs of specimens working in shallow hand-dug trenches. They earn very little for each specimen, despite the hazards of digging in the hot desert sun and breathing huge amounts of rock dust. Most of the profit is made by the middlemen

DOI: 10.1201/9781003288299-5

71

who buy raw specimens from the collectors, then process them in factories in Erfoud. There, hundreds of underpaid fossil preparators use grinding wheels, polishers, micro-sandblasters, and delicate picks to clean up the fossils and make them look good for resale. In many cases, the fossils are too good to be true. It's not unusual for these artisans to "enhance" the fossil by faking parts that are missing, or adding spectacular features like delicate spines, to fetch a higher price. In other cases, lots of individual fossils are cemented into a single slab, where their value is greatly enhanced. At each stage of the process, middlemen and traders add their markup to the specimens, so a fossil which earned the collector a few pennies in Morocco might be worth hundreds or thousands of dollars in the right setting.

There are an estimated 300-400 species of Devonian trilobites in Morocco, many with their spectacular spines and bizarre complex eyes. These are some of the most prized specimens on the market. But by far, the bulk of the fossils are of Devonian cephalopods. These include the incredibly abundant orthoceratid nautiloid, *Temperoceras*, which often occur in dense cluster in black limestones, and are used to make all sorts of objects from decorative wall pieces to sinks, bathtubs, coffee tables, bookends, to candlestick holders (Figure 5.1). Found alongside them in huge numbers are beautiful specimens of ammonoids with the characteristic zig-zag suture, known as goniatites (Figure 5.2). Most of the Moroccan Devonian goniatites are thought to be in the genera such as *Gephyroceras*. With their huge size and elaborate polishing (often highlighting the zig-zag sutures), a good goniatite can fetch a high price. Chances are that if you bought an ammonoid fossil from a commercial vendor, it's a goniatite from the Devonian of Morocco.

Cephalopods in the Age of Fish

The Devonian Period, 419.6–358.8 million years ago, has been nicknamed "The Age of Fish." Even though jawed fish evolved during the previous Silurian Period, it was during the Devonian that fish greatly diversified and developed larger body sizes (up to 10m, or 33ft long) and dermal armor. They became fast as their predatory ability increased, and the once invertebrate-dominated marine food web was forced to adapt (Figures 5.3 and 5.4).

Early plants spread onto land, attracting millipedes from the water, and behind them, scorpions. As time went on, continents shifted and grew, and

Figure 5.1   A slab of Moroccan *Temperoceras* orthocones polished for the fossil trade. Orthocones were washed together after death and preserve the direction of flow. (Photo by Donald Prothero.)

the life surrounding them became more abundant and complex. While life on land was just beginning, life in the ocean was rapidly changing, and the long-lived, plodding nautiluses had to up their ante as well. The world was warmer, and the shorelines of the world were largely concentrated around the Equator—a perfect recipe for shallow seas and shorelines migrating hundreds of miles inland. During the Silurian and Devonian, huge coral reefs blossomed everywhere. Tabulate and rugose corals sprung up in their myriad, alien forms like wildflowers from another world. Large, shallow-water forms such as *Hexagonaria, Pleurodictyum,* and *Favosites* (Figure 5.5) became the Devonian counterparts to today's *Acropora.* Fish dove for the gracefully swaying crinoids with large, coprophagous snails attached, and solitary horn corals like *Heliophyllum* and *Cystiphilloides* climbed to excessive heights, stretching their tentacles as close as they could to the water's surface.

The coral and sponge reefs of the Devonian Period have never been surpassed in their sheer sprawling size. They blanketed countless nautical miles

Figure 5.2 Donald Prothero holding a Moroccan goniatite from the Devonian, polished for the fossil trade.

of ancient shoreline with layer upon layer of coral, sponge, and bryozoan encrustation. Many of these reefs certainly would have dwarfed the Great Barrier Reef, with its area just over 132,000 square miles. Hard corals of the Devonian were fundamentally different from the corals of today, however. Six-sided colonial corals, called "tabulate corals," became abundant, though the first symbiosis between the modern stony corals and colorful algae would not be for several hundred million years. Most of the Devonian reef corals were solitary, occupied by just one anemone-like polyp. Their single-polyp exoskeletons resemble downward-pointed cones, hence their common name "horn corals."

The reefs also provided infinite opportunities for their millions of nektic inhabitants to specialize. Evolutionary innovation accelerated in all species and with it, new relationships between predators and prey. This was an excellent time to be a marine invertebrate, provided your species won the adaptation lottery. Every species had to respond quickly to the rising dominance of armored fish if it had any hope of continuing on (Figure 5.6). If the

Figure 5.3 **Alex and Annabeth Bartholomew marveling at (and facepalming) the incredible diversity of mostly Devonian fossils at the Paleontological Research Institution in Ithaca, NY.**

fish generated necessity, one nautiloid order, Bactritoidea, would become the mother of invention—in the form of two new daughter cephalopod subclasses—Ammonoidea and Coleoidea. Bactritoidea, a straight conical shell or orthocone, is seen as a transitional form between nautiloids and early ammonoids, based principally on the interior structure of the shell.

The primary difference between Bactritoidea and its predecessors was in its siphuncle (Figures 5.6 and 2.8). Throughout cephalopod history, beginning with the plectronocerids, and carried into the modern era by extant *Nautilus* species, the siphuncle had always run through the middle of each chamber and pierced the middle of the septal walls enclosing each chamber. Bactritids had shifted their siphuncle to the ventral wall, along the outer rim of the coiled shell. Stability provided by proximity to the shell's outer wall better supported stronger expulsion of fluid from the chambers of the phragmocone. The siphuncle was also much smaller, and the sutures had changed

Figure 5.4 (a) A Devonian seascape with bryozoans, brachiopods, corals, and early cephalopods. (b) Devonian trilobites, such as this odontopleurid, *Dicranurus*, became extremely specialized at this time as well.

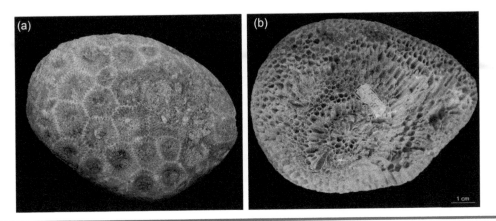

Figure 5.5   Devonian corals: (a) *Hexagonaria* and (b) *Favosites.*

Figure 5.6   Devonian fish chasing after coiled nautiloids.

from the traditional C-shaped suture to a pointed V, allowing more acute control over gases and fluids inside them. At this point, the suture was still concave, as it has always been in nautiloids. Direction in which the septal walls point is the primary diagnostic factor between ammonoids and nautiloids during the overlap in their history.

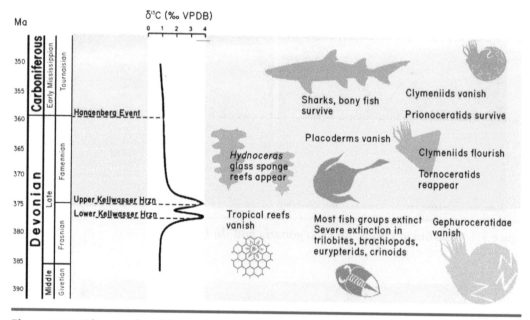

Figure 5.7 **Time scale of the later Devonian extinction events, showing the oceanic carbon isotope curve, and the major extinctions at both the Kellwasser and Hangenberg extinctions.**

Bactritids may have had a slightly more octopus-like soft-part morphology than their ancestors if the nautilus form had not yet been transcended by another nautiloid lineage. They are truly transitional between nautiloids and their two daughter taxa, the coleoids and ammonoids.

Order Agoniatitida: The First Ammonoid

The oceans continued to fill with bizarre, hungry animals, and the pressure to adapt in order to survive mounted (Figure 5.7). In time, the bactritids morphed again. They shifted their siphuncles even further and added new folds to their septal walls. With these changes came the first Ammonoidea, which included the agoniatitids and anarcestids (Figures 5.8). Ammonoids became the dominant cephalopod and stayed the dominant cephalopods from the Devonian Period until the end of the Cretaceous. It is commonly thought that if not for the fateful event which killed the non-avian dinosaurs, octopodes and squid would probably never have spread into anywhere near as many marine ecosystems and ecological niches as they have. The ammonoids were particularly successful in the Devonian because they evaded domination from predators

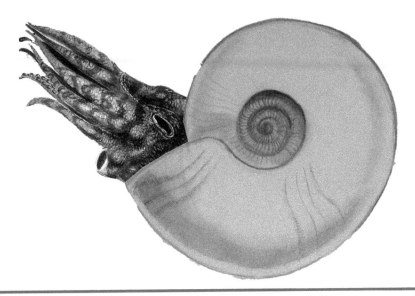

Figure 5.8 *Cabrieroceras* an early agoniatitid ammonoid.

by swimming faster, and simply multiplying faster than predatory fish could eat them all. Both descendants of the bactritids moved into the gaping r-selected niche that had never been occupied by a cephalopod. (r-selected animals do not receive much parental care but have many tiny larvae at once, which increases their odds that just a few will survive and reproduce, even if most are lost.) Even today, a chambered *Nautilus* may live for 20 years, but an octopus—the ammonoids' closest living relative—will in most cases be lucky and very, very old if it lives 2 years. The nautilus, slow in its development and producing relatively few offspring, is considered K-selected. This shift to rapid life cycle and reproduction, called tachytely, is also likely what afforded ammonoids a greater degree of biodiversity of any cephalopod subclass living or past. Ammonoid biodiversity is orders of magnitude greater than that of the nautiloids *or* the coleoids. There are presently 928 known species of coleoids alive on Earth today, and there are about 2,500 species of nautiloids between those living and those in the fossil record. About 5,400 total species of mammals are known, fossil or living. Ammonoid species number between 10 and 20 thousand! There was a time in the history of this planet when shelled, grabby, googly-eyed ammonoids were as ubiquitous as insects.

Agoniatitids are known from Lower to Middle Devonian deposits, and their shells were relatively conservative. They are often neither particularly laterally compressed nor robust, and only lightly ornamented with nodes (Figure 5.9).

Figure 5.9 **(a) Adult and juvenile *Agoniatites nodiferous*, one of the earliest ammonoids. (b) Alex Bartholomew holding up a fragment of a mature *A. nodiferous* from the Cherry Valley Member, Mount Marion Formation.**

The sutures of agoniatitids possess elements of both nautiloids and bactritids, with a subtle nod to the more complicated sutures to come. These early ammonoids have maintained the retrochoanitic siphuncle from their nautiloids ancestors, in which the septal necks point back rather than forward.

Agoniatitids are frequently found in limestones that represent paleoenvironments of moderate turbidity. New York State is home to one of the best localities for huge amounts of agoniatitids—the Cherry Valley Limestone, a member of the Marcellus Shale, which is most commonly known for its central role in hydrofracking controversy. The Cherry Valley Limestone is filled with fossils, and, somewhat uncommon for a Devonian units of North America, it is heavily populated by fossil cephalopods. Ammonoids are so abundant in Devonian sediments of Europe that they have been used there to delineate and divide time. In North America, only Mesozoic ammonoids can be used in this manner due to the relative paucity of Devonian ammonoid localities. Very few agoniatitids are known from the Americas, with none being from South America, and most American agoniatitids coming from the northeastern United States. The Cherry Valley Limestone is special to ammonoid workers because it is a very good exception to the comparatively sparse population of agoniatitid fossils from this part of the world.

The Marcellus Shale produces nautiloids: some are heavily pyritized, as well as the bactritid *Lobobactrites*; and four early ammonoids: *Agoniatites*, *Cabrieroceras*, *Paradiceras*, and *Tornoceras*. *Cabrieroceras* is the lowermost ammonoid occurring in the *Cabrieroceras* Bed of the lower Bakover Shale Member. The next abundant ammonoid found in the Marcellus interval is the large *Agoniatites nodiferous*, found primarily within a thin shale unit called the Lincoln Park Shale just below the main Cherry Valley Limestone. In the Cherry Valley Limestone proper is an abundance of the large *Agoniatites expansus* along with the smaller tornoceratid form known as *Paradiceras*. Finally, in the East Berne Member, directly above the Cherry Valley Member, Tornoceras aff. mesopleuron has been found in the Hudson Valley region of New York State. T. aff. mesopleuron is an important addition to the Eastern North American goniatite fauna as it can be used to help place the base of the Givetian Stage of the Middle Devonian in this region.

Both ammonoids and bactritids persist throughout the Marcellus and up into overlying siltstones and shale of the middle to upper Hamilton Group, demonstrating the answer to one of the most common misconceptions about the way evolution works. When a daughter taxon emerges, many people assume the parent taxon must automatically go extinct. This misunderstanding tends to manifest in the common question "If humans evolved from apes, why are there still apes?" Evolution occurs more like the branches of a tree, and not like a "chain of being" or a "ladder of life". Certainly, a parent does not automatically die the moment they have a child. Humans and apes are twigs on the same branch of the massive tree of life. Likewise, ammonoids branched from bactritids, and both branched from nautiloids, but the times during which members of all three taxa overlapped.

Comparing the geographic distribution of the ammonoids and bactritids at different points in time also provides some clues about the changing marine environment (Figure 5.10) during the Middle Devonian. In New York, *Lobobactrites* and *Agoniatites* disappear above the Hamilton Group, but in European strata, the former persists into the Upper Devonian (by which time, *Agoniatites* had already disappeared in Europe). The habitat became increasingly restricted for *Agoniatites* over the stretch of time when the Hamilton strata were deposited, and they moved west, while habitat of *Lobobactrites* was restricted eastward—and apparently supported them a bit longer.

We have millions of early ammonoids in mid-Devonian marine sediments because Earth was so much warmer in the Devonian Period than it is today. The Cherry Valley Limestone and similar units were formed by transgressive events—rises in sea level often driven by climate change—depositing marine

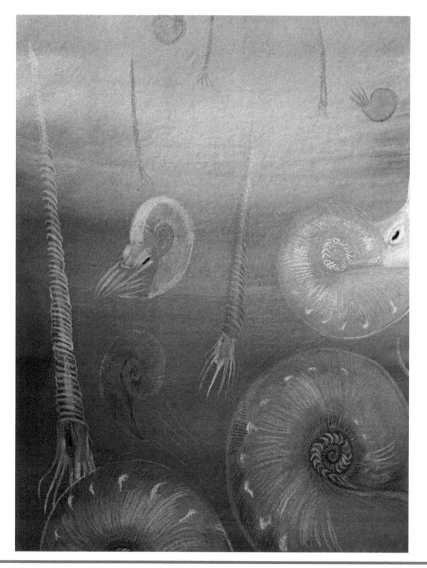

Figure 5.10 **Co-living *Lobobactrites* and *Agoniatites*.**

sediments in the middle of continents. Washing in and depositing thousands
of shells from goniatitids, agoniatitids, and nautiloids during the transgressions, these Devonian limestones are sometimes called "*cephalopodenkalk*,"
a German word that denotes a limestone filled with fossil cephalopods. As
sea level went up, clastic sediments like sands and muds were sequestered in
the nearshore environments and the open water became much clearer and
less muddy. Sunlight was able to penetrate deeper into the water and calcareous algae proliferated on the seafloor, along with small corals like auloporids
(tubular tabulate corals), tiny crinoids, and a few small brachiopods. Mostly,

though, because the water became too deep for larger benthic fauna like larger reef-building rugose and tabulate corals, and bigger brachiopods, we find a concentration of pelagic nekton, especially cephalopods: orthoconic nautiloids, and agoniatitic and goniatitic ammonoids. *Cephalopodenkalk* units are generally not very thick, often just a couple of meters, and tend to be sandwiched between darker, fine-grained clastics like shale.

Agoniatids also differed from nautiloids in their growth and development, and the ammonoids carried this difference with them right up through the end of the Cretaceous. While nautiloids were commonly isometric, changing only in size, and seldom in body proportions, ammonoids frequently changed the shape and ornamentation of their shells. *Agoniatites nodiferous* (Figure 5.11), like many later ammonoids, exhibits very different ornamentation in its juvenile and adult phases. Heavy ribbing, called bullae, dominates in the juvenile stage while just simple rows of small nodes are visible on the adult whorls (Figure 5.9). The juvenile ribs are invisible on adult specimens because new shell coiled over them as the animal grew. This ability to change morphs through the life cycle became a frequent and increasingly more dramatic feature in many ammonoid lineages (Figure 5.12).

Figure 5.11   **Reconstruction of *Agoniatites nodiferous*.**

Figure 5.12 **Agoniatites expansus reconstruction.**

Order Goniatitida

Order Goniatitida shifted their siphuncle all the way to the outer part of the rim of the shell, or the ventral wall. They added surface area to their shallowly lobed sutures by evolving them into full zig-zags and flipped them around to face their convex edge out toward the aperture. Though the siphuncle was still small, the chamber shapes enabled goniatitids to control buoyancy with a proficiency that had never before been attained. This is partly because zig-zag suture afforded ammonoids another game changer: the added structural support from chevron sutures possibly allowed ammonoid shell walls to become paper-thin.

The outward-pointing chevrons coincided with another change. Up to this point in the evolution of ammonoids, the septal necks—structures which support the siphuncle—faced backwards in the chambers (retrochaonitic). In juvenile goniatitids, they retained the retrochaonitic septal neck. However, as the goniatitids reached maturity, a shift occurred, and the septal necks switched to facing forward (Figure 5.13). This orientation, called prochaonitic, allowed the brunt impact of propulsion to be better supported by the siphuncle. It is logical to deduce from this that goniatitids could swim a bit faster than other cephalopods of the same geologic age. Many goniatitids (Figure 5.14) had a combination of both prochaonitic and retrochaonitic phases, but normally, the retrochaonitic phase occurred in early life, and the prochaonitic phase came about as the animal approached adulthood. However, over the evolution of Paleozoic ammonoids, the septal necks eventually came to be prochaonitic throughout ontogeny, even in the juvenile stage (Figure 5.13).

Modern nautiluses are not derived from ammonoids, and as a result, they only possess retrochoanitic necks. When considering that they have endured hundreds of millions of years with septal necks that do not well support fast motion, a question emerges: why didn't nautiloids, from the very first *Plectronoceras*, simply evolve prochaonitic necks in the first place?

Figure 5.13 **Septal neck in goniatitid cross section.**

Figure 5.14 **Illustration of *Goniatites crenistria* from Hooke et al. (1895) and (b) suture of *G. crenistria*.**

The idea that animals' bodies are perfect, or at least, that they are intended to be perfect, is an antiquated one. Truthfully, bodies are not designed to be the absolute best they can be. Evolutionary change is energetically costly and deeply based. Animals' bodies are not perfect, they are only as good as they need to be. The septal neck of early nautiloids was not subjected to nearly as many evolutionary and ecological pressures as the ammonoids; the nekton at the time of *Plectronoceras* was sparser and relatively benign. The septal necks of the first nautiloids grew in the backward-facing direction for reasons we don't know. Perhaps there was some physiological reason for this, which the soft tissue of these animals is no longer available for us to learn from, or perhaps it was arbitrary, a coin flip of evolution. Either way, evolution occurred the way it always does: just as well as it has to, and not an iota better, for the times and conditions in which it finds itself *right now*, with no ability to account for what may occur thousands of years down the road.

Paradise Lost: The Devonian Extinctions

The Late Devonian was the time of the planet's biggest mass extinctions since the end-Ordovician extinction event (Figure 5.7). The waves of extinction struck different groups at different times, at both the end of the Late Devonian Frasnian Stage, and the end of the final stage of the Devonian, the Famennian. The first wave occurred in the Late Frasnian (Kellwasser Event), but this cataclysmic chain of events began earlier in the Givetian Stage, the last stage of the Middle Devonian which immediately precedes the Frasnian. We call the "Frasnian-Famennian Extinction" by the name "Kellwasser

Event," primarily because of where it occurs at the boundary at Kellwassertal in Saxony, Germany. Here, a sharp extinction occurs with biodiversity abruptly cut off at the boundary between the Frasnian and Famennian strata. The Kellwasser Event appears to bear similarities to more infamous extinction events, including the end-Cretaceous extinction and the controversial Anthropocene. Time and time again, the almost half-billion-year-old recurring nightmare for cephalopod biodiversity has been climate change. The same greenhouse atmosphere responsible for the warm, shallow seas that produced all life were suddenly building up and coming to a head. Now, global warming became a liability to everything that once benefited from it. The Kellwasser warming event created warm, low-oxygen conditions in many marine ecosystems which once hosted goniatitids. This is comparable to the "dead zones" created by modern-day climate change. Rock strata that contain the Frasnian-Famennian boundary within them almost never show an upward succession of goniatitid fauna.

The Kellwasser Event hit many types of animals particularly hard, including the ammonoids, trilobites (most of the typical Devonian groups vanished), brachiopods (especially the diverse spiriferids), the armored jawless fish, most of the lobe-finned fish, and jawed placoderm fish, while the snails, clams, and bryozoans escaped with only minor extinctions (Figure 5.7). Animals which survived the smaller extinction events which ended the Frasnian stage suddenly began to dwindle in the Famennian. The elaborate trilobites which thrived in the Givetian Stage of the Middle Devonian, including odontopleurids and Asteropyginae, were gone. Corals were decimated, and their fellow reef builders, the stromatoporoid sponges, disappeared altogether. Just a few goniatitids made it through. The family Gephuroceratidae, which included manticocerids (Figure 5.15) and *Ponticeras* (Figure 5.16), a family of goniatitids which once constituted large portions of the ammonoid fauna in the Late Devonian, suddenly vanished.

Because the manticocerids accounted for a huge percentage of goniatitid populations at this time, the true toll of this extinction on ammonoids puzzled paleontologists for decades. Initially, the extinction was thought to be much more devastating to cephalopods than it truly was. Like *Agoniatites*, *Manticoceras* changed considerably over its growth history (Figure 5.17) and also had a high degree of variation within one species. (We call this "intraspecific variation.") In 2007, paleontologists Dieter Korn and Christian Klug discovered another root cause for the significance of this extinction toll's false inflation.

Figure 5.15 **Manticoceras reconstruction.**

Figure 5.16 **Ponticeras reconstruction.**

In ammonoids from the Jurassic and Cretaceous, misshapen shells are ubiquitous. In the rapidly produced shells of Mesozoic ammonites, keels wander, sutures grow on top of one another, and in heteromorphs, coils can become very disorganized—two members of the same species may look wildly different from each other. An exact reason for the chaotic growth of later ammonoids' shells is not known. It has been assumed for decades that Paleozoic ammonoids exhibited less of this unpredictability in shapes and sizes of members of a single species, called intraspecific variation. We previously discussed how goniatitic sutures exhibit very little variance, with complexity values and overall geometry that almost never show significant change. *Manticoceras* flouted this otherwise seemingly-universal law of goniatitids: a single *Manticoceras* species could show an alarming number of shell shapes and sizes (Figure 5.17).

The eventual recognition of so many different looking ammonoids as one species suggests likely confusion for paleontologists of the past. One way the falsely inflated biodiversity of *Manticoceras* certainly led researchers astray was in the significance of the Kellwasser Crisis itself. When extinction events occur, a primary way their toll is measured by paleontologists is through the number of species lost. Fossil assemblages do not consistently

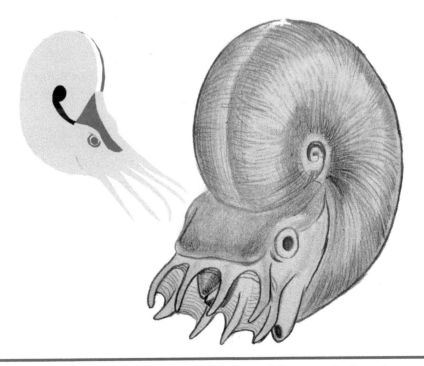

Figure 5.17 *Manticoceras* **intraspecific variation—differences in the robustness.**

represent living populations of animals. With marine or riverine fossils especially, assemblages are often a place where remains of animals were washed or otherwise deposited together over the course of years. Conversely, a sparsely populated fossil assemblage may represent many more individuals than the preservation lottery accounted for. Population dynamics on groups of prehistoric animals are not always possible, and when even the most abrupt extinctions occur, it is simply more reliable to measure their severity in the number of species, not individuals, that are wiped out.

However, the biggest victims of the Kellwasser extinction were the giant coral-stromatoporoid reefs that had dominated since the early Paleozoic. Stromatoporoid sponges vanished completely, while most of the tabulate and rugose (horn) coral genera were wiped out as well, so only a few solitary corals were left to survive into the late Paleozoic. The complete obliteration of the warm-water coral reef assemblage undoubtedly explains why there is so much extinction in many of the groups that lived in and around the reefs, such as the brachiopods and primitive ammonoids.

There is also a strong temperature signal in the extinction. The major group of brachiopods to vanish was the atrypids, which were exclusively tropical in their distribution. Likewise, the tropical reefs of the world were decimated, whereas corals found in deep, colder water were mostly unaffected, and reef communities in the polar latitudes (such as the Parana Basin of South America) also escaped severe extinction. Most striking of all is what organisms replaced the tropical reefs: giant reefs of the glass sponge *Hydnoceras* (Figure 5.18), which are particularly common in the Late Devonian of upstate New York. These were previously known from

Figure 5.18 **The glass sponge *Hydnoceras*, a common Late Devonian reef-building organism from upstate New York.**

the colder, deeper waters and then spread to the shallows, apparently in response to the dramatic cooling of tropical waters.

The signal of tropical cooling can be seen in many other indicators, such as the geochemistry of oxygen in the oceans, which indicates a massive ice buildup. Indeed, there are Late Devonian glacial tills in many places, including the polar regions of Gondwana, and even as close to the ancient tropics as Pennsylvania, Maryland, and West Virginia (Rockwell Formation).

Famennian Stage: The Latest Devonian

The last stage of the Devonian, the Famennian, saw goniatitids bounce back, but not without much resistance. A reliable presence of ammonoids does not return until considerably above the Frasnian-Famennian boundary, indicating that goniatitids took some time to recover. The apparent total absence of ammonoids does not mean they went completely extinct: a single fossil suggests any number of unpreserved individuals of that species. In fact, one of the most significant families of goniatitids which survived the initial Kellwasser extinction, the tornoceratids (Figure 5.19), soon proved to be a family of "Lazarus taxa," appearing to have gone extinct previously, and showing up in the fossil record again only at this stage. (The name "Lazarus" refers to the Bible story where Jesus raises Lazarus from the dead). The most famous example of a Lazarus taxon is the coelacanth, a lobe-finned fish once only known to humans from fossils. Believed for years to have become extinct since the Mesozoic, coelacanths survived in hiding until they were discovered only 84 years ago, when, in 1938, South African natural history

Figure 5.19 **A pyritized *Tornoceras*. (Courtesy James St. John.)**

museum curator Marjorie Courtenay-Latimer learned of a freshly caught coelacanth in the wares of some local fishermen. She was able to confirm the discovery and announce to the world that this fossil fish was actually still alive and well.

Only the tornoceratid *Phoenixites* unabashedly survived without an abrupt disappearance and reappearance. Like the manticocerids, significant work remains to be done on tornocerid systematics, and the number of species confused researchers for a long time. If these ammonoids are simply rapidly speciated descendants of the tornocerid casualties, then no Lazarus phenomenon occurred. Another possible explanation is simply the suppression caused to species when they are forced to compete for common resources, causing a population decrease in a given species that resembles extinction in the (imperfect) fossil record. When competing animals disappeared (whether other ammonoids or something else), the Lazarus taxa re-emerged.

For goniatitids, surviving this event was dependent on the unique characters of individual juveniles. Evolution is an ongoing, perpetual process, and it is at work, however undetectably, in each living thing. As such, some juvenile ammonoids were already beginning to exhibit differences in their shell morphologies, and moreover, developmental innovations, at the moment disaster struck. The protoconch is the shell of a molluscan embryo. Living nautiluses, usually the best proxy for primitive ammonoids, have protoconches that measure almost 10 mm in diameter. Early Devonian goniatitid protoconches could measure nearly 2 mm in diameter. By the Jurassic, ammonoid protoconches were all safely less than 1 mm.

The first decrease in protoconch size for ammonoids started and ended in the Devonian, with most lineages nearly halving the diameter of their embryonic shell. The smaller embryos proved more successful in the low-oxygen waters produced at the end of the Frasnian. The umbilicus of many of these shells boasted new ornamentation which has sometimes been interpreted as functionally advantageous for goniatitids braving these new ocean conditions.

Order Clymeniida

After the Kellwasser Event, another ammonoid order rose up alongside goniatitids during the last stage of the Devonian. Contrary to what many fossil cephalopod lovers believe, clymeniid ammonoids are not goniatitids, and there are some considerable differences between these two ammonoid

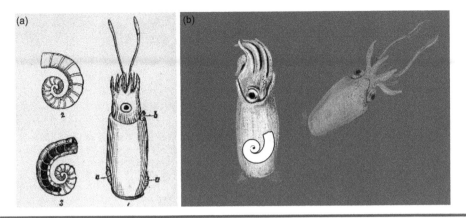

Figure 5.20 (a) *Spirula,* including its internalized gyrocone shell. (b) Spirula in life position with internalized shell in situ.

orders. Unlike any other ammonoids, the siphuncle in clymeniids is positioned along the dorsal wall of the shell instead of the venter, crowding up against the earlier coils in the shell. This may seem like a small difference, but it is reasonable to posit that clymeniids exhibited different movement, behavior, and soft-part mechanisms from goniatitids of their time as a result of the aberrant position of their siphuncle alone. Today, the only cephalopod shell with a dorsal siphuncle is the internalized gyrocone of *Spirula* (Figure 5.20), a small modern squid whose entire body is oriented vertically in resting position. Because the living chamber on many clymeniids was also very narrow, it is likely that they may not have been the best nektonic swimmers. However, their body chambers were usually significantly longer than 180°. Hydrostatic models have indicated that, unlike *Spirula*, experiments have shown that most clymeniids had a similar orientation to a "regular" ammonoid with a ventral siphuncle (Figure 5.21). Even though many of their body chambers were proportionally longer than those of many other ammonoids, clymeniids would likely have had pretty positive buoyancy, and probably floated.

Clymeniids were also somewhat idiosyncratic in their suture geometry. While goniatitids exhibited clear boundaries between their saddles and lobes, clymeniids took this a step further. The lateral lobes of clymeniids are particularly pointed, with some species exhibiting a bent point in the lateral lobe. Clymeniids did not usually have as many umbilical lobes as goniatitids, but this bend in the apex of the lateral lobe potentially serves an interesting purpose. It represents the first time that an ammonoid divided within the lobe itself in order to increase complexity.

Figure 5.21   *Clymenia laevigata* in an orientation similar to life position.

Other than these features, clymeniids were fairly conservative. Some were highly involute, but most are not visibly different from goniatitids until one peered inside of their shells. Enter *Soliclymenia*, generally considered to be one of the bizarrest ammonoids ever to live. The shell of *Soliclymenia paradoxa* (Figure 5.22) was triangular!

Figure 5.22   *Soliclymenia paradoxa* reconstruction.

Hanging with the Hangenberg: The End-Devonian Event

About 359 million years ago, or 12 million years after the Kellwasser Event, another extinction event struck the planet (Figure 5.7). Known as the Hangenberg Event, it killed off about 80% of species on Earth. The toll was not as severe for most types of animals as it had been in the Kellwasser Event, but huge numbers of trilobites and eurypterids, corals, crinoids, and other animals still perished. The event was also detrimental to ammonoids; while enough goniatitids survived to repopulate in the Carboniferous, the end-Devonian event would end up destroying the short-lived clymeniids. The total extinction of clymeniids was probably due less to their internal morphology and likely had something to do with the shape of their shell. Many clymeniids' shells barely widened with each new whorl and many resemble coiled straws in cross section (Figure 5.23). Only the most laterally

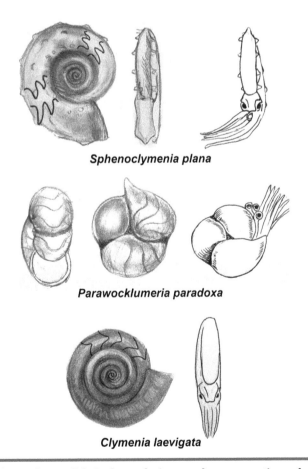

Sphenoclymenia plana

Parawocklumeria paradoxa

Clymenia laevigata

Figure 5.23 **Various clymeniids in lateral view and cross section, showing their lateral compression.**

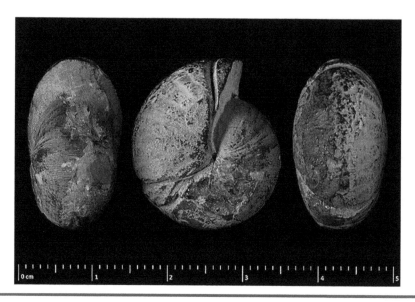

Figure 5.24   ***Prionoceras.***

robust goniatitids, a select few members of the family Prionoceratidae (Figure 5.24), passed through the boundary, and their descendants would become the ceratitids and prolecanitids.

The Hangenberg Event also finished off the armored placoderm fish completely. Out of the huge radiation of fishes in the "Age of Fishes," only sharks, lobe-finned fish, and ray-finned fish survived into the Carboniferous. The Hangenberg Event wiped out 97% of the aquatic vertebrate species, so the only survivors in the Early Carboniferous were sharks less than a meter long, and fish and amphibians less than 10 cm long.

So what caused the two Devonian extinctions? Many ideas have been proposed, from asteroid or comet impacts, to massive volcanic events, to plate tectonic changes, to depletion of the oxygen in the ocean. The evidence for impacts has never been confirmed and doesn't seem to match the pattern of extinctions, and the tectonic events are too slow for such abrupt events, so they are not strongly supported as a cause.

The oceanic oxygen depletion is undoubtedly real, since the typical Upper Devonian deposits (especially during the Kellwasser Event) are anoxic deep-water black shales, such as the Cleveland Shale in Ohio, the Chattanooga Shale in Tennessee, and the New Albany Shale in Illinois. However, many geologists think that the anoxia and the clear signal of global cooling are related, so as the ocean basins became dramatically cooled and stratified, their bottoms were starved in oxygen and deadly for organisms. In times of global cooling, air and water lose energy to move, and they stop mixing, resulting in oxygen depletion in the ocean. Thus, the only well supported causes are dramatic cooling in the tropics, and oceanic stagnation.

What could have caused this to happen? Since 2002, volcanism has been a popular explanation. It turns out that there were gigantic eruptions of basaltic flood lavas at both the Kellwasser and Hangenberg events, mostly in what is now Siberia and Russia. The Viluy lavas in the Siberian craton have been recently redated, and they match the dates for the Kellwasser Event. The Pripyat-Dneiper-Donets large igneous province, just north of the Caspian Sea, has been dated to just before the end of the Devonian, and thus could explain the Hangenberg events. Such huge eruptions could have produced enough sulfur dioxide and dust in the atmosphere to contribute to global cooling, and with it, the Devonian glaciers would grow and sea level would drop. Not only would this help produce stagnant anoxic ocean basins, but the high sulfur content of the gases would also contribute to sulfur contamination of the oceans.

In recent years, a number of geologists have pointed to the fact that the planet developed large-scale forests for the first time in the Late Devonian. Up until this time, only algae and simple, low-growing land plants affected the carbon dioxide balance in the atmosphere. Large trees not only absorb and trap a lot more carbon dioxide, but they also speed up deep weathering of the soil with their roots, which makes the soils absorb carbon dioxide as well. The Earth was not completely done with this greenhouse state yet (as we shall see, it returned for its last hurrah in the Early Carboniferous). The later Devonian was the first significant pulse of global cooling since the latest Ordovician, and it inflicted itself upon a huge tropical biomass which had persisted for millions of years, making it totally unprepared for sudden cooling. The Hangenberg extinction was more severe than any previous extinction in several important ways. Unlike the Ordovician extinction, which wiped out more genera and species but did not fundamentally change the ecological communities, the Late Devonian extinctions wiped out much of the tropical fauna and changed the world by wiping out the reefs, as well as decimating the huge radiation of Devonian fish groups. Beginning in the imminent Carboniferous Period, life of the world's oceans would never look the same.

Future Reading

Algeo, T.J. (1998). Terrestrial-Marine Teleconnections in the Devonian: Links between the Evolution of Land Plants, Weathering Processes, and Marine Anoxic Events. *Philosophical Transactions of the Royal Society B: Biological Sciences*, 353(1365), 113–130.

Bond, D.P.G., Wignall, P. B. (2008). The Role of Sea-Level Change and Marine Anoxia in the Frasnian-Famennian (Late Devonian) Mass Extinction. *Palaeogeography, Palaeoclimatology, Palaeoecology*, 263(3–4), 107–118.

Brannen, P. (2017). *The Ends of the World: Volcanic Apocalypses, Lethal Oceans, and Our Quest to Understand Earth's Past Mass Extinctions*. Ecco, New York.

Hallam, A., Wignall, P.B. (1997). *Mass Extinctions and Their Aftermath*. Oxford University Press.

MacLeod, N. (2015). *The Great Extinctions: What Causes them and How they Shape Life*. Firefly Books, London.

McGhee, G. (1996). *The Late Devonian Mass Extinction: The Frasnian/Famennian Crisis*. Columbia University Press, New York.

McGhee, G. (2013). *When the Invasion of the Land Failed: The Legacy of the Devonian Extinctions*. Columbia University Press, New York.

Miller, A.K. (1938). *Devonian Ammonoids of America*. The Geological Society of America.

Prothero, D.R. (2020). *The Evolving Earth*. Oxford University Press, New York.

Racki, G. (2005). Toward Understanding Late Devonian Global Events: Few Answers, Many Questions; in Jeff O., Jared M., Wignall P. (eds.), *Understanding Late Devonian and Permian-Triassic Biotic and Climatic Events*. Elsevier, London.

Sallan, L.C., Coates, M.I. (2010). End-Devonian Extinction and a Bottleneck in the Early Evolution of Modern Jawed Vertebrates. *Proceedings of the National Academy of Sciences*, 107(22), 10131–10135.

Chapter 6

Survivors: Prolecanitids, Ceratitids, and the Origin of Ammonites

In the previous chapter, we saw how a few lineages of ammonoids survived the Devonian mass extinctions. This is a pattern we see again and again—one major extinction after the next, ammonoid biodiversity was repeatedly decimated, but a few lineages always managed to survive. Soon after, these survivors underwent an explosive adaptive radiation and repopulated the oceans with new ammonoid taxa. The ammonoids did this again at the end of the biggest mass extinction of all at the end of the Permian Period (about 252 Ma), and yet again after the mass extinction that ended the first period of the Age of Dinosaurs, the Triassic Period (around 200 Ma).

When extinction events occur, vacant ecological roles (called niches) are left in the ecosystem. The survival of just a few goniatitid lineages from the Devonian into the next geologic period, the Carboniferous, offered new chances for ammonoids to evolve and innovate. The goniatitids rebounded and flourished again (Figure 6.1), in an ever-constant arms race: first against the arthropods, and then the perpetually increasing resistance from fish. Cephalopods, including ammonoids themselves, were almost certainly a common predator of other ammonoids. At the same time, ecosystems on land were becoming more sophisticated and advanced. The ever-complexifying lives of ammonoids were not disrupted by the first sprigs of moss on dry land, nor by the millipedes who

DOI: 10.1201/9781003288299-6

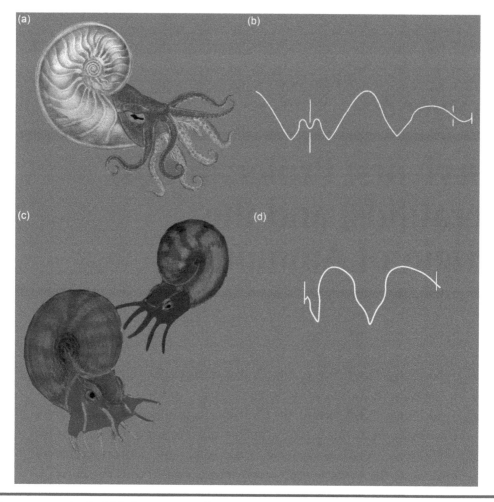

Figure 6.1   Goniatitids became much more diverse throughout the Carboniferous. Reconstructions of (a) *Eumorphoceras girtyi*, (b) suture of *E. girtyi*, (c) *Eosyngastrioceras hesperium*, and (d) suture of *E. hesperium*. (Sutures Courtesy USGS.)

walked out of the sea in search of primitive mosses, nor by the scorpions who walked out of the sea in search of millipedes. By the Carboniferous, these early land ecosystems had grown into lush fern forests.

It was during those golden years of the Paleozoic Era that oxygen reached its highest concentration in our atmosphere (possibly as high as 35%, rather than 21% like today), and gigantic dragonflies called *Meganeura* cruised boldly over the early forests. The oxygen boost was in part a product of enormous expansion of land plants. The beginning of the Permian Period (289.9 million years ago) was marked by more complex vegetation able to live farther from water, and diversifying of land creatures. It was the time of *Dimetrodon*. Life on Earth was evolving rapidly.

Complexity reigned on land and in the sea, and the sutures of goniatitids followed suit. Later goniatitids exhibited new saddles and lobes in their sutures. With even a slight increase in the complexity of their sutures, new variations appeared (Figure 6.2).

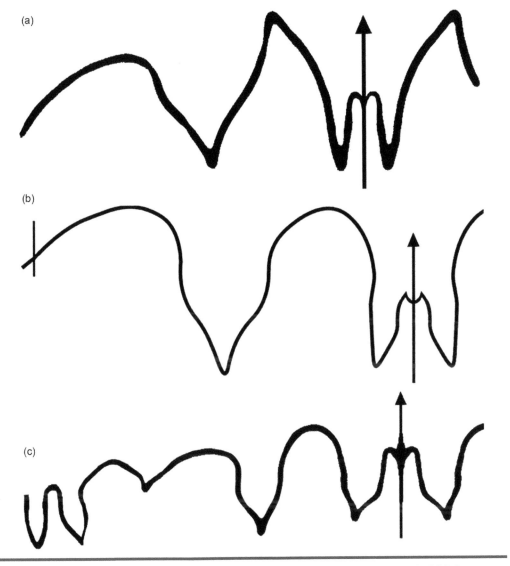

Figure 6.2   Sutures of (a) Devonian goniatitid (Redrawn from Korn and Ebbighausen, 2008), (b) Carboniferous goniatitid (Redrawn from Titus 2015), and (c) Permian goniatitids (Redrawn from Shen et al., 2004).

Order Prolecanitida

While life on land became increasingly complex, life beneath the seas was also thriving, and goniatitids carried on, unaffected as the world sometimes literally shifted around them. It was during the Carboniferous that an elusive order of ammonoids arose alongside the goniatitids. Comparatively little research interest has been focused on the prolecanitids of all the ammonoids. Part of this is due to the relative rarity of their fossils, and the small number of prolecanitid genera. They are frequently dismissed as a transitional form between goniatitids and ceratitids.

However, the value of prolecanitids as a transitional ammonoid order should not be discounted. The internal anatomy of prolecanitid fossils offers subtle changes that many paleontologists consider indicative of radical biological change. Prolecanitid living chambers were much shorter than their goniatitid cousins, causing them to float face upwards in the water column.

Prolecanitids did another thing very differently from their ancestors—they embarked on an accelerated increase in the complexity of their sutures. While they retained simple saddles and lobes from goniatitid ancestors, the prolecanitids added more of them. When lobes increased, their individual size became more restricted. The primitive *Protocanites* (Figure 6.3) had

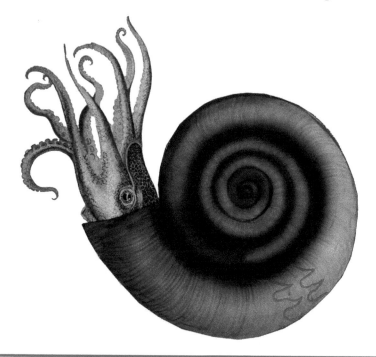

Figure 6.3 *Protocanites* reconstruction with highlighted suture lines.

Figure 6.4 *Daraelites elegans.*

just three or four umbilical lobes in a mature suture, while the much later *Bamyaniceras* could exhibit as many as 15. The appearance of more but smaller saddles and lobes was likely an important departure from goniatitids in the way that these ammonoids lived. Despite this, the sutures are composed of many smooth parabolas stitched together that do not so easily tell their secrets. The trajectory of the prolecanitid suture was, nonetheless, a steep and mostly linear slope of increasing complexity. The changes to the prolecanitid suture were so stark and occurred so rapidly that, unlike the other ammonoid orders, septal geometry by itself has repeatedly been relied upon to recognize genera and species. The dependence on sutures for prolecanitid systematics has been so prevalent that septal geometry was used to trace all of the Mesozoic ammonoids to the prolecanitid genus *Daraelites* (Figure 6.4).

Order Ceratitida

As the world ocean swarmed with goniatitids and prolecanitids in the Late Permian, the lobes of prolecanitids began to divide, but they did not divide completely. This resulted in sutures that had a smooth, arching saddle and a bifurcating or even slightly squiggly lobe (Figure 6.5). From these delicate forms, an unassuming new ammonoid called *Paraceltites* (Figure 6.6) took to the seas. This new ammonoid swam faster and enjoyed even better control over its buoyancy than either of its goniatitid or prolecanitid ancestors. Cruising the oceans with novel anatomical structures, *Paraceltites* was the first ceratitid.

Figure 6.5 **Typical ceratitic suture in** *Protrachyceras pseudoarchelonus.* **(Courtesy Ian Alexander.)**

Ceratitids quickly became ubiquitous, and they dominated the niches held by ammonoids throughout the Late Permian. The early ones shared many of the qualities that enabled certain goniatitids to survive the Devonian. Their shells were robust, and lacked real ornamentation. Ceratitids were conservative and hardy, and they spread rapidly around the world. While Ceratitids were outwardly conservative, their sutures told another story. They had added intricacy to the goniatitic chevron. Arching U-shaped saddles were paired with serrated lobes in *Daraelites* and *Boesites*. The serrations, the first time ammonoids truly subdivided within individual saddles and lobes, resulted in statistically gigantic increases in disparity of ammonoid septa for ceratitids and every ammonoid that came after them. This jump in sutural complexity was matched by a decrease in the thickness of the shell's chamber walls. Universally, ammonoid sutures decrease in complexity as they increase thickness. The cause for this universal trend, which was first noticed in rapid changes in complexity

Figure 6.6 *Paraceltites elegans* reconstruction.

of the prolecanitid lineage, has been argued for over a century. Did the frilliness increase the structural integrity of the shell so that less calcium carbonate would need to be secreted by the animal? Do the diverging patterns correspond to changes in lifestyle or function? The exact reason for these differences may never be known, but the fact that geometric complexity is always directly variable with thinning of the septal wall is not disputed by anyone.

The "Mother of All Mass Extinctions"

For millions of years, ceratitids cohabited the oceans with goniatitids and prolecanitids, but 252 million years ago, disaster struck again: the Paleozoic utopia would be demolished by the largest mass extinction event of all time. Known as the Great Dying, the end-Permian event wiped out nearly all species of marine fauna. Global warming spiked and stayed scorching. Sea level became unstable, likely shutting off major warm- and cold-water currents, and introducing new ones by force.

Life on land suffered a 70% reduction in species and would not bounce back for millions of years. Life in the seas was reduced by more than 95%. The end-Permian extinction finally marked the end of the reign of trilobites, along with the eurypterids, all but a handful of brachiopods, and more than half of all families of animals. The paradise of shallow epicontinental seas of the Paleozoic Era was no more. In the Permian, the bottom sediments of these shallow tropical seas were full of the single-celled fusulinids, whose shells resemble rice grains. These were wiped out at the peak of their evolution during the Permian catastrophe. There was no hope of survival for the rugose and tabulate corals that built the sprawling reefs in the sprawling reefs of the Devonian. In a few million years, coral reefs would vanish from the planet altogether. In the later part of the Early Triassic, these archaic coral groups would be replaced by the scleractinian corals, which apparently evolved from a new group of polyps in the Early Triassic, and today they dominate modern reefs.

All evidence points to unprecedented volcanism which transformed the planet's climate, triggering massive trophic cascades, and choking off life in the shallow marine realm. The Siberian Traps (Figure 6.7) comprise a "Large Igneous Province," an enormous expanse of erupted lavas in northern Russia. With an area about seven million square kilometers (or three million square miles), the Siberian Traps are close in size to the whole of Europe, or close to the size of the continental United States. The volcanos that these rocks represent would have been so massive that even a mild eruption at their scale would have been considered a global event. They continually erupted about four million square kilometers (or 2.48 million square miles) of lava for several hundreds of thousands of years.

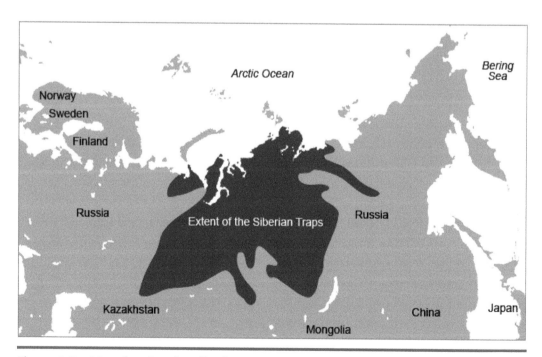

Figure 6.7 **Map showing the Siberian Traps in northern Russia.**

By the time of the faunal turnover, enough magma had been spewed by the Siberian Traps to cover either Europe or the continental US with a kilometer-thick layer.

Such eruptions from deep mantle sources would have released enormous volumes of greenhouse gasses, especially carbon dioxide and sulfur dioxide. In 2013, scientists estimated that these eruptions released 8.5×10^{13} metric tonnes of carbon dioxide, 4.4×10^{11} metric tonnes of carbon monoxide, 7.0×10^{12} metric tonnes of hydrogen sulfide, and 6.8×10^{13} metric tonnes of sulfur dioxide—staggering amounts of volatiles gasses that completely transformed the atmosphere into a super-greenhouse full of toxic sulfates and sulfides, which quickly changed ocean chemistry. Making the conditions even worse is that the lavas ignited thick seams of Carboniferous coal that lay in the Siberian bedrock, which burned intensely. Just like humans burning coal in the past two centuries, these burning coals released enormous volumes of carbon dioxide and other greenhouse gasses.

The oceans then became supersaturated with carbon dioxide, making them too hot and acidic for many species, and killing nearly everything that lived in them. Ocean temperatures are estimated to have reached over 40°C (104°F), far hotter than most tropical marine life can even stand temporarily. While carbon dioxide can rapidly produce a greenhouse state when enough of it concentrates in the atmosphere, methane is an even more potent greenhouse gas than carbon dioxide. The warming of the seafloor may have released immense quantities of frozen methane from the bottom sediments, producing a massive inrush of methane. There are many places where the black shales and other geochemical evidence suggest that the waters became depleted in oxygen and maybe even poisoned by hydrogen sulfide. The atmosphere was also low in oxygen and full of excess carbon dioxide, so land animals above a certain size nearly all vanished, and only a few smaller lineages of synapsids, reptiles, amphibians, and other land creatures made it through the hellish planet of the latest Permian, and survived to the aftermath world of the earliest Triassic.

Still, the fact that the onset of the Siberian Trap eruptions predate the Great Dying by up to three hundred thousand years suggests that a single cataclysmic eruption is not to blame for the disappearance of the extinct species. Monstrous volcanism was already well underway at the point of disappearance for many of the Permian casualties, slowing deposition only slightly and continuing on for about half a million years after it claimed them.

For reasons unrelated to their impending doom, ceratitid and goniatitid biodiversity had been in steady, albeit somewhat gradual, decline for much of the Permian. In the wake of the cataclysm, about 3% of all ammonoids made it past the Permian boundary event. This gave ammonoid recovery in the Triassic what appeared to be a snowball's chance in hell, and furthermore, it supports the consensus among researchers that the ammonoids' pinnacle of biodiversity in the Mesozoic spawned from a single Permian ancestor. Though the survivors belonged to only three genera, their descendants belonged, overwhelmingly, to a single genus.

A Blank Slate

These lone survivors faced an empty Triassic ocean but, provided that they could survive in it, virtually boundless opportunities to specialize in the vacant niches. It was not just ammonoids that were decimated; the oceans were also void of the fiercest predators of ammonoids. With unchecked claims on the oceanic real estate, a biological spring was quickly set into motion after the ammonoids' darkest winter. *How* quickly this happened was almost unbelievable to researchers, who believed for most of the twentieth century that this rebound stretched out over at least ten million years.

The real absurdity of ammonoids' survival is that of the three genera (Figure 6.8) that survived the Permian and lived into the Mesozoic (*Otoceras*, *Episageceras*, and *Xenodiscus*), only one—*Xenodiscus*—would speciate so rapidly and with such diversity that little more than a million years later, the oceans were once again filled with at least as much ammonoid biodiversity as they had been right before the Great Dying. The rapid replacement of ammonoid biodiversity appeared unrealistic to paleontologists for decades. Initially, scientists argued that a period of at least 20 or 30 million years was necessary to bring ammonoids back to their Late Paleozoic numbers. But over time, this figure reduced to about 5 million years. In 2009, however, evidence from a team of several laboratories across France and Switzerland showed that in just about 1 million years, ammonoid biodiversity had completely rebounded. All ammonites in the later Triassic, Jurassic, and Cretaceous are traceable back to the descendant genus of the xenodiscids, *Ophiceras* (Figure 6.8b).

A good place to see the explosive diversification of ceratitids just above the Permo-Triassic boundary is in Union Wash in the Inyo Mountains, just east of the Owens Valley and Sierra Nevada in California (Figure 6.9a). The

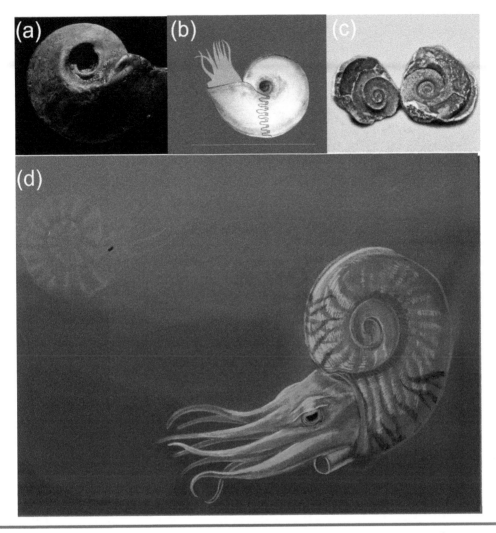

Figure 6.8 (a) **The three surviving ammonoids:** *Otoceras, Episageceras,* **and** *Xenodiscus.* **(b)** *Ophiceras,* **right, and** *Xenodiscus,* **left, floating in a vacant sea.**

section of limestones dips sharply to the east, flanking both walls of the wash. As you drive east, you climb into younger and younger beds. The locality is not easy to reach: a long drive up a rough dirt road climbing the alluvial fan debris filling the canyon, until you reach the end of the road and park. Then it's a hard climb up a steep wash until you reach the fossiliferous layers, then a scramble up to reach them and begin collecting. Driving up the wash, you see thick deposits of reddish Lone Pine Formation, which is full of typical Permian fossils, such as the fusulinid foraminifera (single-celled protistans that secreted shells shaped like a grain of rice), plus brachiopods, bryozoans, corals, crinoids, and mollusks. Then you cross the

Figure 6.9 **The Triassic Union Wash *Meekoceras* locality. (a) The exposures of Permian and Triassic beds in Union, Washington. (b) Deformed specimens of *Meekoceras* from the ammonoid-rich limestones. (Photos by D.R. Prothero.)**

Permian-Triassic boundary, and the limestone above it are Lower Triassic—and almost devoid of megascopic fossils. But sticking out of the walls of the wash are prominent ridges formed from particularly resistant limestone beds. As you clamber around those ridges, you look closely and see that they are chock-full of tiny ammonoids, about the size of a Lifesaver candy. They are packed in enormous numbers in certain beds, so much so that some hand samples have dozens in them. When you find really well-preserved specimens with the outer shell dissolved off, you can see their distinctive sutures with the U-shaped saddles and lobes. These specimens all belong to the well-known Early Triassic ceratite genus *Meekoceras*. In some of the resistant limestone ridges, the tectonic forces that have tilted these rocks to almost vertical have also sheared and deformed them. The ammonoid shells from these layers have also been sheared and deformed (Figure 6.9b), so their shells are no longer round and circular, but often stretched out into an oval shape. Such deformed fossils are useful to geologists who are analyzing the structural forces that deformed the rocks. They can measure the strain direction on specimens like these and calculate which way the forces were applied that tilted the rocks long ago.

Just two million years after the Permian-Triassic event, the number of ammonoid genera had nearly doubled. No one could anticipate just how quickly ammonoids were able to recover from the violent die-off of all the goniatitids and nearly all the ceratitids. Within the first five million years of the Triassic, there were over 200 genera of ceratitids living in the oceans.

Unlike their goniatitic relatives from the Devonian, ceratitids were not restricted to individual localities: instead, their territories were large lateral swaths determined by latitude, a trend in distribution that would stay with many ammonoid lineages throughout the Jurassic and Cretaceous. The emptying of the oceans allowed for these rapidly breeding and rapidly evolving ammonoids to spread out, capitalizing on a relaxed planktonic larval phase free of vulnerability to predators; enjoying their own excellent buoyancy and good, old-fashioned ocean currents to scatter them like spider babies ballooning on their parachutes of silk.

The ceratitids which won the extinction lottery had a few things in their favor that allowed them to rebound in the earliest Triassic when so many other ammonoids perished. The shapes of the surviving ceratitid shells were measured using Raup's coiling morphospace (Figure 3.3): W for width, D for total cone length ("distance from the generating curve"), and S for aperture shape, which is expressed as a height-to-width ratio. Amazingly, the W and D values of the ceratitids which survived match almost exactly the same W and D values as the goniatitids which survived the Hangenberg Event. Even more surprising than the shared robusticity of ammonoids from the two extinctions, the S values (which describe aperture or whorl cross-section) of Triassic ceratitids were *nothing* like those of the end-Devonian survivors. The (S, D) coordinates of end-Permian ammonites matched the distribution of their later descendants in the world's next mass extinction at the end of the Triassic.

The Earth itself was in the throes of change, with the tectonic beginnings of Pangaea's demise. As the Triassic wore on, North America began to pull away from its supercontinent neighbors, South America and Africa. Evidence for the sudden tectonics can be found in rift valleys in the involved continents and in rapidly deposited terrestrial sediment at these intervals.

This tectonic activity would generate brand new paleoshorelines; it meant that local sea level on any continental slope even remotely involved would be subject to change, and with it, the jurisdiction of the ammonoids. Relative or local sea level (RSL) is fundamentally different from global sea level: changes are often due to changes in tectonics and/or sedimentation, which means it tends to occur independently of eustatic, or global sea level change (GSL). GSL is most often caused by rising or falling worldwide climatic conditions, such as ice volume on the poles. Changes in local sea level can oppose global trends, canceling them out in one area and resulting in environments that are changing in the opposite trajectory of everything around them.

Ammonoids were arguably affected by both global and local sea level changes, but always more immediately and viscerally by changes in local sea level. In much of what is now the Middle East, the rifting of Pangaea caused incredibly unstable conditions of sea level rise and fall, despite the fact that the entire world was undergoing a steady eustatic regression—worldwide sea level fall brought on by global cooling.

A Fresh New Look

The fairly plain shells of the Paleozoic ammonoids were gone in the Mesozoic, and ammonoid shells began to become heavily ornamented (Figure 6.10). The smooth surfaces and bold, shallow ridges of the Permian ceratitids were replaced on many shells by an outer layer of fine, but deep, ribbing. Tubercles emerged, and sometimes, hundreds of nodes saturated the shell exterior in a halftone pattern. Sometimes called "micro-ornamentation" (although they were clearly visible to the naked eye), these fine-resolution patterns would stay with ammonoids throughout the remainder of their tenure on Earth. Many ammonites were also striped, boasting fine color patterns called striations which followed the axis of coiling. While raised keels (ventral edges) only occurred in a few Paleozoic lineages, the keel shape of many Jurassic ammonoids became more pronounced.

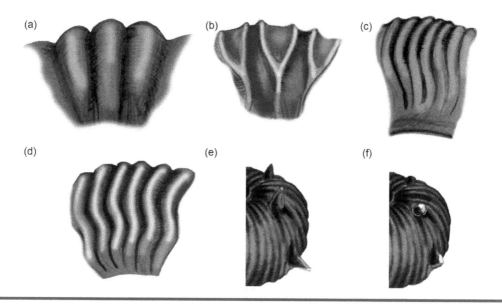

Figure 6.10 **Common types of ribs (bullae) in Mesozoic ammonoids: (a) plicate, (b) primary-secondary, (c) sinuous, (d) biconcave, (e) spines (tubercles), and (f) nodes.**

There could be several evolutionary advantages for micro-ornamentations. Richard Cowen of the University of California Davis argued that extremely textured shells acted as visual confusion to predators—essentially, a camouflage act. Peter Ward of the University of Washington, noting how thin ammonite shells became by the Mesozoic, examined whether the forking ribs of Jurassic and Cretaceous ammonites added to the tensile strength of the shell. Geerat Vermeij of the University of California Davis suggested, based on texturing as a defense mechanism in land snails, that the new ornamentation in Triassic ammonoids may have counteracted the bite pressure of predatory fish. Vermeij has called it the "Mesozoic marine revolution" because most shelled invertebrates of the Triassic had to cope with a wide range of new predators with shell-crushing teeth, especially among the fish and marine reptiles. Many mollusks, such as clams and some snails, survived by becoming good at burrowing; others, like scallops, learned to swim away by clapping their shells together like castanets and jet-propelling themselves as the water in their mantle cavity was expelled when their shells closed. But most Triassic clams and snails survived the Mesozoic marine revolution by making their shells thicker and more ornamented, so they are harder for shell crushers to get a grip on and break. Some of the ornamentation of Triassic ammonoids might also be explained this way.

As one of the fastest animals to bounce back, for several million years ammonoids had very few natural predators. But they were not the only creatures to experience an overhaul of their lineage at this time: the same opportunity was extended to nearly all the species who survived the Great Dying.

Tabulate and rugose corals, which had amassed unmatched reef area in the Paleozoic, vanished completely during the end-Permian extinction, and there were no coral reefs anywhere in the Early Triassic. But eventually they were replaced by a new type of colony, the scleractinian corals or hexacorals, the ancestors of all of today's corals. These apparently evolved from polyps which did not secrete a hard calcareous corallite, so they didn't fossilize well. Scleractinians grow slowly, dependent on the symbiotic algae that they host in their tissues to sustain them.

The hunters of ammonoids were given a similar opportunity to rebuild, and new superpredators ascended to the apex of the ocean ecosystem. This meant that by the Middle Triassic, ammonoids that had any hope of survival in the Mesozoic would need to make additional changes. Two lineages of fish: chondrichthyes, the cartilaginous fish that include rays and sharks, and osteichthyes, the bony fishes, which comprise most of the fish we see today, experienced similar booming populations in the empty canvas that was the earliest Triassic ocean. While bizarre, archaic sharks from the Paleozoic like *Helicoprion* (Figure 6.11) and *Xenacanthus* held on for a little while in the Early Triassic, sharks and rays in particular are believed to have enjoyed an explosion at this time close to that of the ammonoids, although fossilization of cartilaginous fish is difficult and less successful than the preservation of bony fish. The primitive chondrichthyes, ancestors of all living sharks today, had weathered the sea change in deep water where the effects of climate change could not be felt. When conditions became suitable, they resurfaced.

However, sharks were far from the only predators ammonites had to contend with. In the approximately 130 million years between the Great Dying and the moment *Tiktaalik* first set flipper on land, reptiles had become established, flourishing in the terrestrial ecosystem. About two million years after the Permian mass extinction, they had taken to the sea in the form of early ichthyosaurs (Figure 6.12), dolphin-shaped reptiles that swam in the open ocean gave live birth and, inevitably, ate smaller animals—including fish and cephalopods.

Figure 6.11   *Helicoprion* hunting ceratitids.

Figure 6.12 **A mother *Chaohusaurus* and her baby hunting early ammonites.**

Phylloceratoidea and the First True Ammonite

From the Middle Triassic on, the vast majority of ammonoids would share a common ancestor that originated at this time. This ancestor is almost certainly a member of the suborder Phylloceratoidea, as this is the group that dominated throughout the Triassic. By the onset of the Jurassic Period, in a staggering 50 million years, the Phylloceratoidea would diversify into four families.

Phylloceratids took everything that the Triassic ceratitids had done, and did it better. The serrations filled both the saddles and lobes, creating the first ammonitic suture. These ammonitic sutures were far from the complicated forms that had yet to evolve, however. With scalloped arches in all directions, they more closely resembled the oakleaf collages of the Fauvist painter Henri Matisse than either the goniatitic chevron or the late Cretaceous bifid labyrinth. (Their name comes from the Greek *phyllon*, or leaf.) Still, considering the path from the ceratitic suture to the insane complexity of later forms, the phylloceratid oakleaf seems like a reasonable transitional stage.

[To many enthusiasts and students, the words "ammonite" and "ammonoid" are interchangeable, but there is a difference between the two terms. Ammonoids are a subclass of animals—a category which ranks below class (in this case, Cephalopoda) and above the order (for ammonoids, the orders are Agoniatitida, Goniatitida, Clymeniida, Prolecanitida, Ceratitida, and the true ammonites—Ammonitida). Therefore, ammonoids include any order within the subclass, and ammonites are simply the last of those orders, which are typified by highly ornate complex florid sutures. We will call them "ammonites" when we are only discussing the order Ammonitida. Phylloceratids and their many descendants can therefore safely be called "ammonites."]

The appearance of phylloceratids also heralded the days of ammonite shells that were frequently thinner than an eggshell. While it is easy to think of ammonites as either heavy, solid rocks or like thick-shelled nautiluses, neither image is really accurate. By the time of the first ammonites, the ammonoid design had dramatically thinned its shell, with many ammonites under 10 cm in diameter adopting a paper-thin shell. The primary function of the shell was no longer solely for protection, it was for buoyancy, and a balloon offers much more assistance in this area than a brick.

Though the great increase in geometric complexity of the sutures (Figure 6.13) meant the phylloceratid lineage had much thinner septal walls than any previous ammonoids, the relationship between the thickness of the outer shell wall and the sutural complexity is less agreed upon by paleontologists. The thinness of the septa may be just a function of their newfound frilliness.

Even in the Early Triassic, ammonites were about the most rapidly evolving marine megafossil in the world. Some of the first familiar ammonites, like *Meekoceras* (Figure 6.14a), *Ussurites*, and *Monophyllites* (Figure 6.14b), dominate Smithian (earliest Triassic) sedimentary rock layers.

Another Mass Extinction

If ammonites were exploding in population and biodiversity in the beginning of the Triassic, their ubiquity had become incredible by the end of the period. Phylloceratids filled the oceans, taking on a vital role as a food source to countless fish and reptiles. But just as it had happened so many times before in the history of ammonoids, an interruption to this prosperity was imminent. The end of the Triassic meant that ammonoids would need

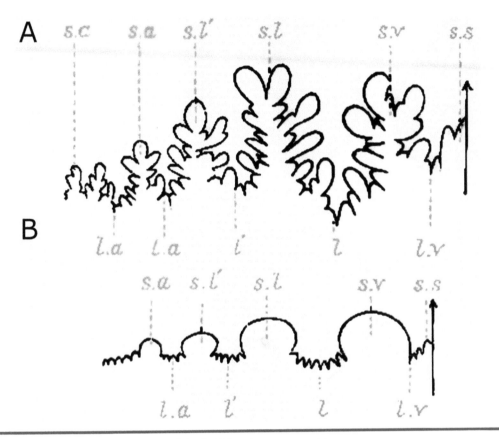

Figure 6.13   Suture patterns of Early Triassic ammonoids.

to survive yet another mass extinction event. Though none would be as devastating as the end-Permian event, more than 75% of all marine life was wiped out, and the fortunate ammonites that survived would need specific traits to endure both the disaster and, inevitably, what lay beyond it.

Of the "Big 5" major mass extinction events first recognized by David Raup and Jack Sepkoski in 1982, the fourth or fifth largest is the one at the end of the Triassic, about 201 Ma. It is only slightly smaller than the end-Cretaceous event that wiped out the dinosaurs, with about 35% of marine genera and possibly 50% of marine species vanishing, and an equally severe crash in the land animals as well (at least 42% of the terrestrial vertebrates, according to one estimate).

The details of the Late Triassic extinctions are complicated, because the last stage of the Triassic (the Rhaetian Stage) has a poor fossil record in many parts of the world, so it is unclear whether many animals died out at the very end of the Triassic, or near the end of the Triassic, at the Norian-Rhaetian boundary. As Tanner and colleagues pointed out in 2004,

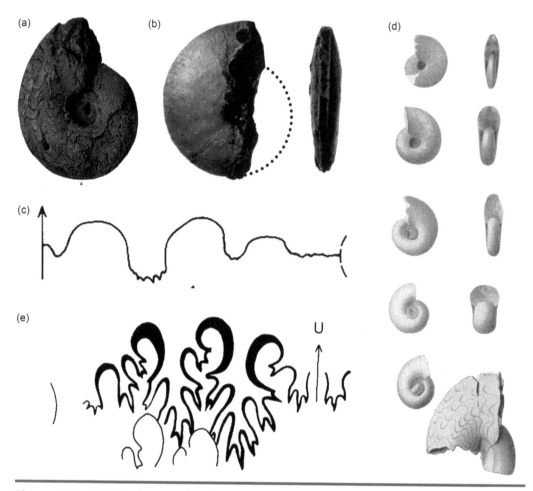

Figure 6.14 **(a)** *Meekoceras* **fossil with sutures, (b)** *Meekoceras* **fossil, (c)** *Meekoceras* **suture, (d)** *Meekoceras* **intraspecific variation, and (e)** *Monophyllites* **suture.**

some of the things that were thought to have vanished at the end of the Triassic now appear to have been more impacted by the end-Norian extinctions. In any case, most scientists see the Triassic extinctions as a series of severe events, just like there were two pulses in the Ordovician extinctions, the Devonian extinctions, and the Permian extinction.

In the marine realm, the most obvious victims of this extinction were the ceratitic ammonoids, which had been the dominant shelled cephalopods of the entire Triassic. They had a first wave of extinction at the end of the Norian Stage but vanished completely at the end of Rhaetian. A few phylloceratid ammonites made it through, and through them, another bottleneck radiation would occur in the Jurassic. These survivors radiated into an evolutionary explosion of new lineages in the Early Jurassic that would comprise the Ammonitida.

Other mollusks, such as the clams and oysters, suffered as well, but marine snails did not seem to have been affected as much. The brachiopods that had survived the Permian extinction were hit again. Spiriferids, a group of brachiopods which had been dominant in the Devonian and Carboniferous, and barely survived the Permian extinction, finally vanished in the Early Jurassic after being decimated in the Triassic extinctions. Even the coral reefs, which had been destroyed in the Permian, then finally recovered in the Middle Triassic, were wiped off the face of the Earth again and did not return again until the Sinemurian, the second stage of the Early Jurassic. These reefs were much more similar to the reefs of today than they were to the reefs of the Devonian.

The fossil record of bony fish is not complete enough to tell which events caused much extinction among them. However, another group of marine vertebrates, the jawless eel-like conodonts, which had survived all the extinction events of the Paleozoic, met their final demise at the end of the Triassic. Ichthyosaurs, a reptile occupying the ecological niche held today by dolphins, had become the dominant marine reptile during the Triassic. A large portion of primitive ichthyosaurs went extinct in the end-Triassic event, but would then recover and evolve into more advanced groups during the Jurassic.

In the terrestrial realm, there were some dramatic changes in the dominant vertebrate groups. The Early Triassic was dominated by huge protomammals (formerly called "mammal-like reptiles") and primitive relatives of crocodiles, as well as gigantic flat-bodied amphibians. By the end of the Triassic, all three of these groups faded out and vanished, leaving the world to the dinosaurs. The giant amphibians were replaced by tiny frogs and salamanders, and the protomammals were replaced by the first true mammals in the Late Triassic, which were tiny shrew-sized insectivores. The first pterosaurs also evolved in the Late Triassic and soon came to rule the skies of the Mesozoic.

So what caused the Late Triassic extinctions? As was the case with many other mass extinction events, a wide variety of culprits have been blamed. After the 1980 publication of the end-Cretaceous asteroid impact model, many scientists assumed that there would also be signs of an impact at the end of the Triassic. One of the prime candidates was the huge Manicouagan impact crater in Quebec, which is 100 km (62 miles) across, the biggest impact event known during the Mesozoic, and second only to the end-Cretaceous Chicxulub crater in size. Naturally, in 1987, a lot of scientists were pointing at it as the Triassic killer. Interestingly, ammonites close to the crater at the Triassic-Jurassic boundary seem to have died out instantaneously, whereas the same taxa held on a little longer as one looked further away from the impact site.

However, the Manicouagan crater turned out not to be the culprit at the end of the Triassic. In 1992, the crater was precisely dated, and its age was 214 Ma—far too young to have anything to do with the extinctions at the end of the Norian (227 Ma) and far too old to have anything to do with the end of the Triassic (201 Ma). Today, the non-effects (for organisms other than ammonoids) of this huge impact event stand as an object lesson of how not to get ahead of the data, or jump to conclusions about impacts always causing extinctions. Other, smaller impacts have been dated closer to the end of the Triassic, but they are too small to cause all the extinctions of the entire Late Triassic interval. Gradual sea level drops were also blamed for the extinctions. But the fact that the extinction was far more severe in the land realm, unaffected by any changes in sea level, and in the sea, was more heavily felt by ammonoids than by other taxa, seems to rule that idea out.

Instead, the likeliest culprit is (once again) volcanism and the associated climate change, the same killer we saw at the Permian extinction (Siberian lavas), the Devonian extinction (Viluyer lavas), the late Ordovician (Pripyat-Dneiper-Donets lavas), and as we shall see in Chapter 10, the end-Cretaceous extinctions (Deccan lavas). During the Late Triassic, huge rift volcanos called the Central Atlantic Magmatic Province, or CAMP, erupted as North America and South America ripped away from Africa as Pangea split apart and the North Atlantic began to open. The first pulses of eruption ran from Nova Scotia to Morocco at the end of the Norian and climaxed through the Rhaetian with the biggest pulse right at the end of the Triassic. Eventually it formed a blanket of 11 million square kilometers of lava, with the volume estimated at three million cubic kilometers of basaltic rock. Based on the latest radiometric dating, the eruptive pulses spanned at least 600,000 years. The flows eventually spanned the area from Brazil to Scandinavia, making them the most widespread eruptions in Earth history. The Moroccan lavas alone formed a layer over 300 m (1,000 ft) thick.

Like the other mass extinctions mentioned already, the main killer from such huge basaltic lava flows over a wide landscape (known as "flood basalts") would be the gigantic volumes of greenhouse gasses, especially carbon dioxide and sulfur dioxide. Most of the volcanic rocks like the Palisades Sill along the west shore of the Hudson River northwest of New York City and the Watchung lava flows in northern New Jersey in the Newark Basin are part of the CAMP. Only eruptions on this scale explain the carbon isotope signal found in marine rocks. Such a huge volume of carbon would have made the oceans more acidic, as happened in the Permian. As in the case of some of the Paleozoic extinctions, the most vulnerable organisms were those with easily dissolved shells made of aragonite ("mother of pearl") such as the goniatitic ammonoids, or those with little direct control of their mineralization of their calcite skeletons (such as corals and calcified sponges).

The ammonoids have proven to be hardy survivors. They were hit with global mass extinction events in the Late Devonian, the Permian, and the Triassic, yet a few lineages survived and then underwent explosive evolution within a few million years of the catastrophes that nearly wiped them out. In the Jurassic and Cretaceous, ammonoids achieved their heyday, with multiple rapidly evolving lineages of ammonites.

Further Reading

Benton, M.J. (2003). *When Life Nearly Died: The Greatest Mass Extinction of All Time*. Thames & Hudson, London.

Benton, M.J., Twitchett, R.J. (2003). How to Kill (Almost) All Life: The End-Permian Extinction Event. *Trends in Ecology & Evolution*, 18(7), 358–365.

Berner, R.A. (2002). Examination of Hypotheses for the Permo-Triassic Boundary Extinction by Carbon Cycle Modeling. *Proceedings of the National Academy of Sciences*, 99(7), 4172–4177.

Brannen, P. (2017). *The Ends of the World: Volcanic Apocalypses, Lethal Oceans, and Our Quest to Understand Earth's Past Mass Extinctions*. Ecco, New York.

Brayard, A., Escarguel, G., Bucher, H., Monnet, C., Brühwiler, T., Goudemand, N., Galfetti, T., Guex, J. (2009). Good Genes and Good Luck: Ammonoid Diversity and the End-Permian Mass Extinction. *Science*, 325(5944), 1118. doi: 10.1126/science.1174638.

Erwin, D.H. (1990). The End-Permian Mass Extinction. *Annual Review of Ecology and Systematics*, 21, 69–91.

Erwin, D. (2006). *Extinction: How Life on Earth Nearly Ended 250 Million Years Ago*. Princeton University Press, Princeton, NJ.

Fox, C.P., Whiteside, J.H., Olsen, P.E., Cui, X., Summons, R.E., Idiz, E., Grice, K. (2022). Two-Pronged Kill Mechanism at the End-Triassic Mass Extinction. *Geology*, 50, 448–453.

Hallam, A., Wignall, P.B. (1997). *Mass Extinctions and Their Aftermath*. Oxford University Press.

Knoll, A.H., Bambach, R.K., Canfield, R.E., Grotzinger, J.P. (1996). Comparative Earth History and Late Permian Mass Extinctions. *Science*, 273, 452–457.

MacLeod, N. (2015). *The Great Extinctions: What Causes them and How they Shape Life*. Firefly Books, London.

McGowan, A.J. (2005). Ammonoid Recovery from the Late Permian Mass Extinction Event. *Comptes Rendus Palevol*, 4(6–7), 517–530. doi: 10.1016/j.crpv.2005.02.004.

Ogdena, D.E., Sleep, N.H. (2011). Explosive Eruption of Coal and Basalt and the End-Permian Mass Extinction. *Proceedings of the National Academy of Sciences of the United States of America*, 109(1), 59–62.

Simms, M.J., Ruffell, A.H. (1989). Synchroneity of Climatic Change and Extinctions in the Late Triassic. *Geology*, 17(3), 265–268.

Simms, M.J., Ruffell, A.H. (1990). Climatic and Biotic Change in the Late Triassic. *Journal of the Geological Society*, 147(2), 321–327.

Smith, J.P. (1932). Lower Triassic Ammonoids of North America. US geological survey professional paper 167.

Tanner, L.H., Hubert, J.F., et al. (2001). Stability of Atmospheric CO_2 Levels across the Triassic/Jurassic Boundary. *Nature*, 411(6838), 675–677.

Tanner, L.H., Lucas, S.G., Chapman, M.G. (2004). Assessing the Record and Causes of Late Triassic Extinctions. *Earth-Science Reviews*, 65(1–2), 103–139.

Whiteside, J.H., Olsen, P.E., Eglinton, T., Brookfield, M.E., Sambrotto, R.N. (2010). Compound-Specific Carbon Isotopes from Earth's Largest Flood Basalt Eruptions Directly Linked to the End-Triassic Mass Extinction. *PNAS*, 107(15), 6721–6725.

Wignall, P.B., Sun, Y., et al. (2009). Volcanism, Mass Extinction, and Carbon Isotope Fluctuations in the Middle Permian of China. *Science*, 324(5931), 1179–1182.

Wignall, P.B., Twitchett, R.J. (1996). Oceanic Anoxia and the End Permian Mass Extinction. *Science*, 272(5265), 1155–1158.

Wignall, P.B., Twitchett, R.J. (2002). Extent, Duration, and Nature of the Permian-Triassic Superanoxic Event. *Geological Society of America Special Papers*, 356, 395–413.

Chapter 7

Jurassic Seapark

Ammonite Pavements

Ancient seafloor bedding surfaces covered with hundreds of ammonites are not unusual in the Lower Jurassic rocks of Europe. One of the most spectacular is the Dalle aux Ammonites ("Slab of Ammonites") in the foothills of the French Alps in Digne-les-Bains region of Provence in southeastern France. (A famous section of the slab is on display at the Kamaishi Museum in Japan.) The site is about 1.5 km north of Digne in the Haute-Provence Nature Reserve. Here, the rocks are steeply tilted by the uplift of the nearby Alps, so the bedding plane is inclined 60° from horizontal. The rocks are Lower Jurassic (Sinemurian) in age, about 191–199 Ma. In one particular outcrop (Figure 7.1a–c), the bedding surface is covered by over 1,500 ammonite shells, some over 70 cm in diameter, interestingly, the unit consists of primarily just one species: *Coroniceras multi-costatum*. The exposed area of ammonites is currently over 350 square meters in area. Multiple studies of the slab have suggested that this was a gradual death assemblage of ammonites that slowly accumulated on the deep-sea bottom over long periods of time, in relatively calm waters with weak currents and slow sedimentation to gently bury them and spread them out. There are burrows indicating that many kinds of invertebrates lived in the muds surrounding the shells, now mostly internal casts, as they accumulated. This, of course, indicates there was no shortage of oxygen in the depositional environment.

In fact, Lower Jurassic ammonite pavements have been known for over a century. In 1847 Lombardy, now a region of northern Italy, a huge expanse of limestone was first described. The limestone had a nodular structure and, unusually for a limestone, its matrix was bright red. Amazingly, this Toarcian

DOI: 10.1201/9781003288299-7

Figure 7.1 *Coroniceras multicostatum* in Dalle aux Ammonites. (a) Panorama of the entire exposure. (b, c) Close-ups of the surface covered by ammonites. (Courtesy Wikimedia Commons.)

Figure 7.2 (a) Ammonites in the Ammonitico Rosso, Southern Alps, Veneto, Val Brenta. (b) Ammonites (*Aspidoceras* sp.) exposed on stratal surfaces in a quarry near the outcrop. (Courtesy Wikimedia Commons.)

(Lower Jurassic) unit was so saturated with ammonites that it is actually harder to find places where it is *not* covered by a myriad of fragmented and whole ammonites (Figure 7.2). A favorite collecting locality for countless Jurassic enthusiasts and geological researchers of several disciplines, the lithofacies is called "Ammonitico Rosso," which, in Italian, simply means "red ammonitic."

It soon became apparent that Ammonitico Rosso stretched far into the Italian peninsula, and it was a correlative layer between the Alps and the Apennines. Slight differences existed between the various localities in which it was found: some layers are not red, but green or pink, and some layers don't even contain ammonites, despite their name. The original locality became known as Ammonitico Rosso Lombardo, and in fair Verona, where there lay some shells, it was called Ammonitico Rosso Veronese. The name of the village Sassorosso, in Tuscany, literally translates to "red stone." The locations had most things in common, nonetheless: argillaceous (clay-filled) matrix preserved, mostly, aragonitic casts of ammonites rather than the original shell. (Shells may recrystallize, but are often partially preserved as calcite.)

It turned out that the Ammonitico Rosso expanded far beyond Italy. The unit has been described in the Polish Carpathian Basin and as far away as Crimea. The unit is called the Geresce Marble in Hungary and has other names throughout Eastern Europe. These massive red, clay-saturated limestones are more significant, even, than they are unusual. It represented a swath of the now-extinct Tethys Sea, an ancient ocean which formed in the Mesozoic during the separation of Gondwana and Laurasia, and ran from the Straits of Gibraltar across the Mediterranean through what would become the Arabian Peninsula, and all the way to Indonesia (India had not yet collided with Asia).

More remarkable than its saturation with color is its saturation with ammonites. As the rifting occurred, these Jurassic marine sediments were pushed into deep water, often with an active carbonate platform being drowned alongside them. As the carbonate-rich sediments were moved around continuously by burrowing invertebrates through a process known as *bioturbation*, nodular clumps occurred.

The sediments descended to depths up to 300 m during the rifting of the Tethys Sea, and the red color of the hematite-clay matrix is caused by oxidation, or rust of the iron in the rock, and indicates that even at that depth, the water was well oxygenated. In Eastern Europe, it is common for not one but two distinct Ammonitico Rosso horizons to be present within the same stratigraphic section. The quantity of ammonites in the Lower Jurassic represented by Ammonitico Rosso is staggering, but their preservation was so close to the time of their death that in many locations, the aptychi of the ammonites are still present. In other localities, the aptychi are preserved alone. While aptychi are uncommon relative to other ammonite fossils, in some cases, entire rocks from the Ammonitico Rosso are filled with them.

The Diversification of Jurassic Ammonites

Like their ceratitid cousins, with whom the early ammonites coexisted for a period, phylloceratid ammonites inherited every ecological role that an ammonoid could occupy. Early in the Jurassic, the genus *Phylloceras* appeared, and from this came the full family tree of later Jurassic and Cretaceous ammonites (Figure 7.3).

During the Mesozoic, the open seas were filled with large reptiles. Mammals were still tiny shrew-like creatures on land; the ancestors of whales would not even dip a hoof into the ocean for another 150 million years, and of course,

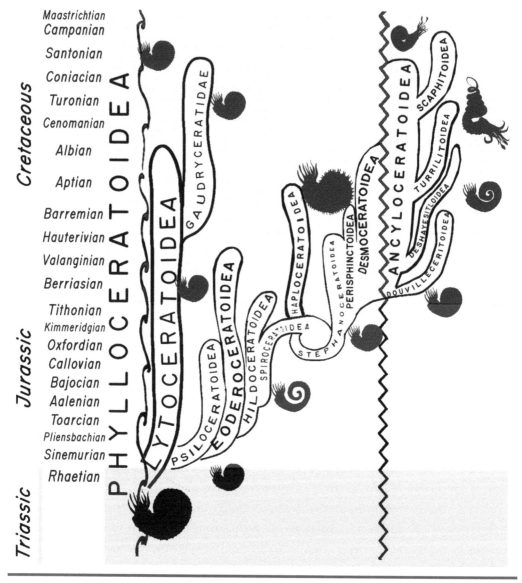

Figure 7.3   **Phylogenetic tree of ammonites based on Yacobucci (2015) in the style of Arkell.**

humans were not present to overharvest the beasts. Plesiosaurs soared through the waters of the Jurassic, their long necks guiding them balletically forth in search of fish who traveled in frightfully fast schools. Alongside them, a rich variety of spirals drifted by, differing wildly in their spiral shapes and ornamentation. These were some of the first true ammonites; they had the thinnest shells and the most complicated sutures up to this point.

It is very unlikely that Jurassic ammonites' thin shells did very much to protect them. Ammonite shells became less and less a defense mechanism, and increased in their adaptations for other abilities. Christina Ifrim realized that tubercles and spines growing along the shell may not have served the obvious purpose of intimidation, but acted as auditory structures which helped ammonites sense vibrations through the water. As the shell grew, the old tubercles would move away from the animal's head, so new spines would appear, creating rows of large ornaments on the shell.

Like an egg, the inner wall of ammonite shells may have been able to heal the animal should the shell be cracked or bitten. However, only certain bites could be healed. The mantle of ammonoids was responsible for new shell secretion, so a shell puncture over the living chamber stood a chance of healing. A bite to the phragmocone had no chance, and the animal would lose control and die. It is almost certain that many bites were fatal to ammonoids, but structural damage of a shell that appears to have glued itself back together is not uncommon in Jurassic and Cretaceous specimens.

Jurassic ammonites became exceptional index fossils, surpassing even the ammonoids of the Triassic. Millions of their fossils litter the sedimentary rocks of this period. Individual species became widespread across thousands of miles and sometimes, multiple continents. Several broadly continuous rock formations attract thousands of researchers and citizen scientists, and "fossil tourists" each year.

Sadly, paleoecological research on Jurassic ammonites is chronically lacking, with most "paleoecological" research on them actually being more descriptive of biostratigraphy and fossil assemblages. Since at least the 1940s, paleontologists have attempted to correlate specific lithofacies (rock types and the paleoenvironmental conditions they represent) to shell morphotypes. In the 1960s, Bernhard Ziegler was the first to assign depth ranges to ammonites based on the rocks in which they were found. Similar ideas were visited in increasing detail by Westermann in the 1990s, with the introduction of the Westermann Morphospace, and Hewitt in 2000. Some of Ziegler's assignments are still accepted, but over the decades, additional interpretations based on them have proven less accurate. Within a single ecosystem, several ammonites may have had different depth distributions in the water column, with some just above the seafloor and others in the open water, or just below the surface. It is possible for all of these ammonites, regardless of life habit, to become preserved together: the fossil record buries shells from all water depths side-by-side, and does not preserve an animal's position in the water column in ways that are obvious just by looking at them.

A commonly employed method of determining an ammonite's place in the water column is to examine whether it had neutral buoyancy. Most ammonites had relatively neutral buoyancy, but may have skewed toward being slightly negative (denser than water, so they would sink). The proportion of the body chamber, which would have been filled with the animal's squid-like soft tissue, to its phragmocone, which would have filled with air, is the best way to tell this. Ammonites with longidome body chambers may have had almost as much of their volume occupied by soft tissue as by air, and they would have been heavier than a brevidome (short body chamber) ammonite, which would have floated more easily. How long a living chamber an ammonite had is just one factor in determining its buoyancy. In his many studies of the living chambered nautilus, American paleontologist Peter Ward determined that only the last three chambers were able to fill with fluid, leaving the most internal spaces in the phragmocone empty and dry. In contrast, water can enter the entire internalized gyrocone of the *Spirula* squid. The capacity any given ammonite had for cameral fluid might have dramatically influenced its buoyancy.

Ammonites' relationship to hydrostatic neutrality was determined in other, less easily knowable ways than their body weight. Peter Ward determined that living *Nautilus* only exchange air and liquid in their three most adoral (closest to aperture) chambers, and use the inner chambers only for gas. *Spirula* squid, by contrast, can bring cameral fluid into all of their chambers. Ammonites show greater diversity in the size of their living chamber than living cephalopods, and we have no way to determine how much of their phragmocone could be emptied of air. According to German researcher René Hoffmann, in spite of some researchers' expectations, a longidome ammonite could *potentially* be positively buoyant, and a brevidome ammonite could make itself quite heavy, depending on other factors.

Today, with the analysis of oxygen and carbon isotopes in the minerals of their shells, we are able to see more clearly where in the water column ammonites lived. In surface waters, lighter weight oxygen isotopes, O-16, are more easily evaporated, leaving a greater percentage of the heavier oxygen isotope, O-18. Organic carbon isotopes are on the opposite continuum: as organic matter sinks to the bottom of the water, C-13 accumulates there.

Carbon and oxygen are both components of calcium carbonate ($CaCO_3$), the inorganic mineral which mollusks metabolize, dissolve, and deposit into new shell material. If the ammonites have been living in surface waters, there would be less O-16 in their shell. If they lived in bottom waters, there will be a lower percentage of O-18, and more C-13. There are other clues as well. We can measure the chemistry of oysters and other bottom-dwelling mollusks to get reading of what the chemistry of the sea bottom was like. If the ammonites found with these mollusks have similar isotopic chemistry, then they are thought to have floated near the bottom as well. If, on the other hand, the shell chemistry is very different from seafloor mollusks, with a high percentage of O-18 and a low percentage of C-13, then it is considered good evidence they lived higher in the water column, probably in shallow water.

Phylloceratoidea

Phylloceratids blossomed from the earliest Jurassic, presumably from their type genus *Phylloceras* (Figure 7.4a), and some interesting new shell morphotypes appeared in the genera *Calliphylloceras* (Figure 7.4b), *Tragophylloceras* (Figure 7.4c), and *Holcophylloceras* (Figure 7.4d).

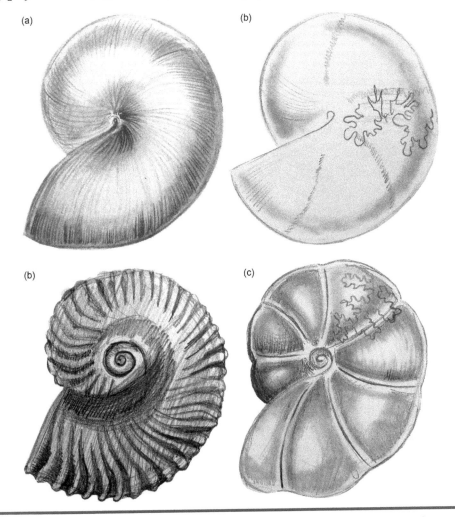

Figure 7.4 **Typical phylloceratoid ammonites: (a)** *Phylloceras,* **(b)** *Calliphylloceras,* **(c)** *Tragophylloceras,* **and (d)** *Holcophylloceras.*

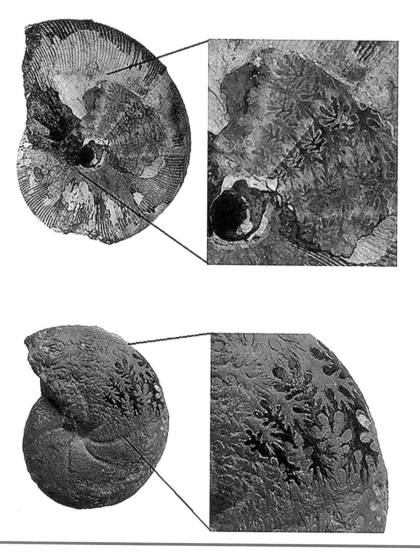

Figure 7.5 *Phylloceras* (a) and *Calliphylloceras* (b) with sutures, both showing a high number of umbilical lobes. (Courtesy Wikimedia Commons.)

The sutures of the phylloceratids were relatively comparable to one another, but their shell shapes could be vastly different (Figures 7.4 and 7.5). The tightness of the coil varied, and *Calliphylloceras* and *Holcophylloceras* developed grooves which appeared intermittently throughout ontogeny. This unique ornamentation is known as radial lirae. Bernhard Ziegler placed phylloceratids in the deepest habitats available to ammonites, at the far edge of the continental slope.

She Sells Seashells by the Seashore

One of the pioneers in the history of paleontology was Mary Anning (1799–1847). She was the daughter of a poor cabinet maker in the beach town of

Figure 7.6 **The only known portrait of Mary Anning, with her little dog Trey, who accompanied her on her fossil hunting expeditions. (Courtesy Wikimedia Commons.)**

Lyme Regis, on the Dorset coast of England (Figure 7.6). Their lives were a tale of tragedy, typical of poor English country folks of the time. Out of ten children, only Mary and her older brother Joseph survived early childhood.

The eldest daughter died in a fire at age 4. Mary herself nearly died as a baby, while being held by a neighbor, Elizabeth Haskins, and lightning struck a tree overhead. Elizabeth and two other women standing under the tree were killed, but amazingly, Mary was unscathed. As Mary's father struggled to make a living by building cabinets, he taught Mary and her brother to collect fossils from the crumbling cliffs east and west of the harbor. Mary and her family sold these fossils to tourists, and she became the subject (or victim) of the famous tongue twister "She Sells Seashells by the Seashore." By far most common fossils were "snakestones" or Lower Jurassic ammonites like *Dactylioceras*, which the Annings collected in abundance and sold to beachgoers as a way to make ends meet. These, along with "Devil's Toenails" (a coiled oyster known as *Gryphaea*), "Devil's Fingers" (fossils of the internal shell of a squid-like creature known as belemnites), "verteberries" (vertebrae), and "bezoar stones" (fossilized digestive stones swallowed by prehistoric animals), were their primary merchandise in the little shop on the ground floor of their home. But when her father died from injuries due to a fall from a cliff in 1810, and her brother went to find a steadier line of work, Mary kept at her lonely quest on the seaside cliffs, collecting not only ammonites, but also the first ever marine reptiles known as ichthyosaurs, plesiosaurs, and the first pterosaurs (or flying reptiles) outside Germany. Her amazing finds of these giant sea monsters opened up new worlds of extinct creatures to the scientific gentlemen of the time.

In the early nineteenth century, the expanding world of paleontology was only open to wealthy gentlemen. Because she was a commoner, a woman, and also not an Anglican, these rich gentlemen refused to admit her to their lofty company of scientific societies. Some of them bought her fossils and published them without even giving her a share of credit. Eventually, however, the naturalists who supported Anning and appreciated her talents managed to raise a small pension for her and give her due credit for her work. Sadly, she died at the relatively young age of 47 of breast cancer, before she ever received the accolades she was due for most of her greatest achievements as one of the most important pioneers of paleontology. Her scientific admirers commissioned a stained-glass window in her honor in the local church, a fitting tribute to one of the most important pioneers in the history of paleontology.

A visit to Lyme Regis today evokes the images of Mary, climbing the cliffs to the east of the harbor. But even more dramatic is the walk west of the harbor, where at low tide, the bedrock is exposed along Monmouth Beach. Here the entire bedrock beneath the shoreline is built of hard Blue Lias shales chock full of ammonites (Figure 7.7). If you take this memorable walk, you can see literally hundreds of ammonites exposed in the eroded bedrock, with various cross-sections visible as they are slowly eroded away. The rock itself is too hard and too brittle to try to break out these fossils, so every visitor gets to see them in all their glory and not worry about the

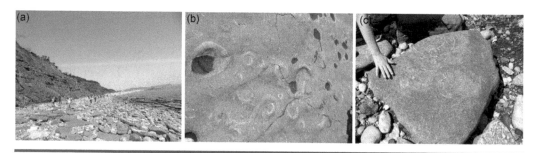

Figure 7.7 **The fossiliferous Lower Jurassic rocks of Monmouth Beach, west of Lyme Regis. (a) Lower Jurassic strata and exposed beach rock at low tide, looking east back at the Lyme Regis harbor. (b) Exposed bedrock at low tide, showing many ammonites in cross section, usually with their central whorls eroded away and filled with water. (c) Close-up of a typical slab of the Blue Lias bedrock, with many closely packed ammonites in cross section. (Photos by D.R. Prothero.)**

locality being mined away and cleaned out. Some of the shells are huge, almost a meter across, while others are smaller. Many have only the outer few whorls exposed, while the inner part of the shell is missing. This is a puzzling fact, although the most popular explanation was that the newly buried shells had their living chambers full of sediment, which withstood the pressure of burial, while the inner gas-filled chambers did not, so they were crushed and eventually dissolved away. Either way, a pilgrimage to Monmouth Beach at Lyme Regis is one of the truly inspiring experiences for professional paleontologists and amateur fossil collectors alike.

Important Jurassic Ammonite Families and Superfamilies

While ammonites in the Jurassic enjoyed unprecedented abundance and geographic distribution, many of the most common genera belonged to just a handful of families and superfamilies. These include phylloceratids, dactylioceratids, lytoceratids, and perisphinctids.

[Family-level names in zoology always end with "-idae," e.g., Dactylioceratidae, with a capital D, and members of the family are informally referred to as "-ids," e.g., "dactylioceratids," with a lower-case d. Superfamilies in zoology often end in "-oidea," e.g., Lytoceratoidea, and their members are usually referred to as "-oids." With type taxa, the family and superfamily may have the same name: Lytoceratoidea is a superfamily, but Lytoceratidae is a family. Lytoceratoidea is a broader term, and Lytoceratidae only describes a subset of them. The same is true of perisphinctoids and perisphinctids, phylloceratoids and phylloceratids, and numerous other ammonoid taxa.]

Dactylioceratidae

Dactylioceras, the type genus of this family (Figure 7.8), was one of the most common ammonites that Mary Anning collected, and today, it continues to be one of the most common fossils in both the Holzmaden shales in Germany, and along the Jurassic Coast. *Dactylioceras* lived in the early part of the Jurassic, generally between the later Pleinsbachian and the early Toarcian stages. It had an exceptionally long living chamber, often comprising more than a full revolution around the shell. Researchers assume that members of this genus were unable to swim very fast, and

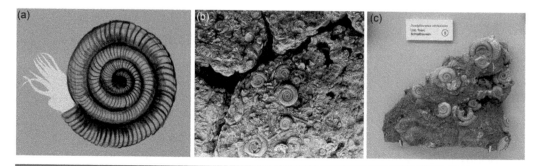

Figure 7.8 (a) *Dactylioceras* reconstruction. There are two species of *Dactylioceras* that are considered particularly common: (b) *Dactylioceras commune* and (c) *Dactylioceras athleticum.*

probably lived near the ocean bottom. Species were large and small and varied both in their ornamentation and their relative abundance.

Dactylioceras is tremendously abundant in many Lower Jurassic units. Its most common species, *D. commune* and *D. athleticum*, are usually less than 10 cm in diameter, and their living chamber could extend more than 360° around the shell. This probably meant that their fleshy tissues were slightly heavier than the suspension that their phragmocone could afford, and they likely lived closer to the sea floor than an ammonite with a short body chamber. Helmut Keupp determined that some *Dactylioceras* specimens exhibited forma aegra verticata, a pathology given to them by seafloor-dwelling crustaceans.

Porpoceras

Porpoceras are larger than *Dactylioceras*, often 30 cm or more in diameter. Though they share their cousins' long-living chambers, and likewise were not very buoyant, they were ornamented with long spines (Figure 7.9).

Holzmaden

One of the most famous Jurassic ammonite localities in the world is part of a black to dark gray shale formation called the *Posidonienschiefer* in German, or Posidonia Shale, which stretches from Germany to the Netherlands, Austria, Switzerland, and Luxembourg. The unit is dated about 183–181Ma, within the Early Jurassic Toarcian Stage, about

Figure 7.9 *Porpoceras* reconstruction.

182.7–174.1 Ma. Although it crops out over a wide region, the most famous fossil quarries in the unit are found near the Bavarian town of Holzmaden. It is a classic black shale, typical of a major episode of anoxia on the ocean floor, and is so rich in organic carbon that it was mined for oil shale in the past. It gets its other name from the fossil clam *Posidonia*, which is especially abundant in these beds. Recently, however, the names of these Jurassic units have been revised, so it is now officially called the Sachrang Formation. Generations of paleontologists, however, have always known it as the Posidonia Shale.

The Posidonia Shale is considered a Lagerstätte, or a "mother lode" of exceptional fossil preservation. The fossils found in the Posidonia Shale include perfectly preserved crinoids, ichthyosaurs (sometimes preserved with their complete body outline, and with fetuses still inside the pregnant females) (Figure 7.10), the exquisitely rare full-body preservation of sharks, often with stomach contents, and hundreds of plant species, many of which are identifiable from more than pollen or spores. The most common ammonite in the Posidonia is *Dactylioceras commune*, a small serpenticone that can be found there by the thousands. *Hildoceras* (Figures 7.9 and 7.10) is also very common. However, almost the full gamut of Jurassic ammonite families is represented, including numerous lytoceratids and phylloceratids.

Figure 7.10   **Ichythyosaur fossil with *Hildoceras*.**

Hildoceratidae

The hildoceratids are one of the most diverse families of ammonites in the Jurassic. A sister taxon of dactylioceratids via a wayward branch of the Eoderoceratoidea, they include St. Hilda's mythologized "snakestones," *Hildoceras* and *Hildaites*, as well as *Harpoceras, Esericeras, Bouleiceras, Grammoceras*, and *Protogrammoceras*, as well as many others. Their sutures are relatively simple, but the sizes and styles of ornamentation for hildoceratids can be incredibly varied. The peristome, or the right and left edges surrounding the aperture, tends to be sinuous or zig-zag shaped. Sexual dimorphism in hildoceratids is common: in most invertebrates, including ammonites and modern cephalopods, females are the larger sex. (Though we don't know for sure, we assume that ammonite females were also larger than males. In some species, the increase in size is accompanied by a wider body chamber, which many interpret to be extra space for an egg pouch.)

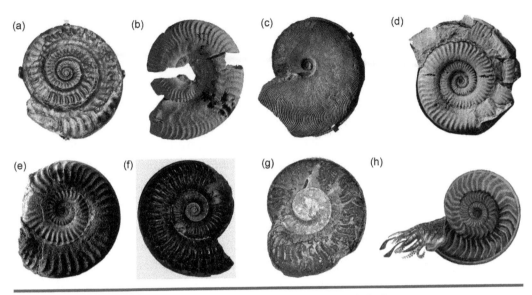

Figure 7.11   An array of hildoceratid ammonite fossils and their suture patterns:
(a) *Hildoceras*, **(b)** *Hildaites*, **(c)** *Harpoceras*, **(d)** *Esericeras*, **(e)** *Bouleiceras*, **(f)**
Grammoceras, **(g)** *Protogrammoceras*, **and (h)** *Hildoceras bifrons* reconstruction.

The rifting that shaped the Tethys Sea also shaped members of the genus
Hildoceras (Figure 7.11). According to René Hoffmann, the changing environments may have played a role in limiting some Jurassic ammonites' size.
Usually characterized by a deep groove alone either side of their keep,
Hildoceras ornamentation and, especially their size, changed relative to
the effects of tectonics in their specific locality. *Harpoceras* was also sexually dimorphic, with males (microconchs) measuring up to about 5 cm, and
females (macroconchs) growing over 40 cm. These distinctively textured
ammonites were probably nektonic, favoring shallow water and contributing
greatly to the food supply of the Jurassic aquascape.

Volga River

The Volga River fauna, from western Russia, includes the giant perisphinctid *Speetoniceras* as well as some of the earliest true heteromorphs, especially *Auduliceras*. Smaller ammonites—*Kosmoceras* and *Kepplerites*—also
appear, but the truly unique thing about the preservation at the Volga River
is the frequency with which ammonites are pyritized. Smaller ammonites
are sometimes completely golden, while larger ammonites are usually just
lightly glittered with it. Pyrite, or fool's gold (made of iron sulfide, or FeS_2),

can affect any fossil or anything buried in the ground, especially underwater. Once an organic object is buried, sulfate-reducing bacteria metabolize oxygen, iron, and surrounding sulfur, forming iron sulfide, which is just the chemical name for pyrite. Pyrite may or may not be visible, as it rusts easily if exposed to humidity and does not always exhibit a submetallic luster. While it increases the interest in ammonites (provided proper precautions are taken to protect these beautiful specimens from humidity), for larger fossils, pyritization can be a problem, infamously dubbed "pyrite disease" by insiders. Many an articulated dinosaur skeleton display has suddenly collapsed or "exploded" because its preparators were unaware of the sneaky pyritization that had gone on inside its bones, which then began to rust (Figure 7.12).

Pseudogrammoceras (Figure 7.13) are a common hildoceratid in Toarcian sediments of Western Europe. They are also host to a quite common and curious pathology. Called the "wandering keel," the outer edge of the shell grows around a small kink, then readjusts and goes back to normal growth. Reasons for this pathology are not well understood, as it occurs in many, but nowhere near all, specimens of *Pseudogrammoceras* and may indicate a common life change or healing from a failed predation attempt.

Perisphinctoids, and their slightly older cousins the stephanoceratids, are both superfamilies which are distantly descended from the superfamily Lytoceratoidea. Like the hildoceratids, they are descended from the

Figure 7.12 **Pyritized *Kepplerites* from the Volga River.**

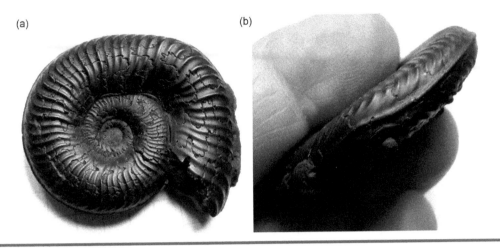

(a)　　　　　　　　　　　　　　　(b)

Figure 7.13　*Pseudogrammoceras* exhibiting the "wandering keel" pathology. (Photos courtesy of Chantal Mars.)

psiloceratoids, the eoderoceratoids, and ultimately, the psiloceratid branch of the ammonite family tree. They were often close to serpenticone in shell shape, although they ranged in their coiling, and a telltale feature of many was the presence of bifurcated ribs on the outside of the shell. Perisphinctids and stephanoceratids were highly prolific and diverse in shape and environment, ranging from under 10 cm to almost a meter in diameter. They were so successful that eventually, they would become heavily relied upon to determine the elusive boundary between the Jurassic and Cretaceous. Ziegler suggested that perisphinctids, stephanoceratids, and similar types of ammonites lived higher on the continental slope, nearer the shore. We know today that there were no true "near-shore" ammonites: though some evidence suggests a slippery outer covering to some ammonite shells that would make it hard to grab them up, their shells were far too fragile to risk being seen by a shoreline predator, an unfortunately placed rock, or a breaking wave.

The vast majority of ammonite sutures increase in complexity over their growth from a juvenile. But sutures from ammonites in the family Perisphinctidae do not. Mature sutures are about as complex as immature sutures on average, but from one suture to the next, the degree of irregularity is greater than just about any other group of ammonoids.

Perisphinctids, like the hildoceratids, can exhibit sexual dimorphism, with massive females and tiny males. Sometimes, males grew lappets, or long protrusions on either side of the peristome. Though protection of animal's soft tissues is one possibility, the role of lappets was probably not to impress females but, due to their small size, to indicate to females that they were reproductively mature males—not immature males, and not immature females.

Figure 7.14 *Pectinatites,* **an ammonite which has shown several examples of changing sex throughout ontogeny.**

It is also possible for individual perisphinctoids to exhibit traits of both male and female traits are various stages of ontogeny. This suggests that from one life stage to another, some ammonites were temporarily male or temporarily female. Ammonoids, like living octopodes and squid, were gonochoristic: distinctly and permanently male or female, unlike many snails, which have both male and female gonads (Figure 7.14).

Stretching from the earliest goniatitid to the latest Cretaceous heteromorph, perisphinctids had distinct stages in the type of ornamentation. Some Jurassic ammonites exhibit distinctly male or distinctly female ornamentation. Less commonly, these types of ornamentation alternate between male and female in an individual perisphinctid. Because the soft-tissue record is so poor in ammonites, it has not been possible to examine the reproductive structures of specimens that alternate between male and female ornamentation. Even the best preserved ammonites often lack many of their organs. Quite possibly, the shell exhibited traits of two sexes, but the soft tissue may have been strictly male or female. However, the two

Figure 7.15 *Peltoceras.*

fossils exhibiting both male and female secondary sex characteristics are specimens of *Peltoceras* (Figure 7.15), an aspidoceratid in the superfamily Perisphinctoidea. In the fossil record, just a handful of specimens with the ability to alternate sex could potentially represent many more living individuals like themselves. Conversely, the low sample size of ammonites whose sex apparently changed ontogenetically could be a simple case of pathology.

Speetoniceras versicolor (Figure 7.16) converged on *Dactylioceras* in all areas except size. Just by looking at their primary-secondary bifurcated ribs, one can tell that these are perisphinctoids. However, like other serpenticones, they had living chambers that extended more than 360° around their shell.

Perisphinctes virguloides (Figure 7.17) is abundant in deposits around the world, but especially in Madagascar. Throughout the Mesozoic, the global ocean salinity was more highly concentrated along the equator and less so at the poles than it is today. During this same time, the Madagascan plate underwent an almost perfectly vertical southward migration through these tectonically parabolic levels of saltiness (Figure 7.17).

Whether due to salinity or another force, preservation of *Perisphinctes* is interesting. They frequently make their way into the commercial fossil trade,

Figure 7.16   *Speetoniceras versicolor.*

5 cm

Figure 7.17   *Perisphinctes virguloides* **are often sliced at the peristome by the commercial fossil trade, making it hard to know how many are preserved with aptychi.**

and *in situ* aptychi are relatively common, making it possible for just about anyone to own a very impressive fossil specimen.

What is remarkable about the aptychi of *Perisphinctes* is that they appear to cover almost the entire area of the shell aperture, and they are extremely thick.

Kosmoceratidae

Kosmoceratids are common in the Volga River fauna, as well as elsewhere across Eurasia. The shells of *Kosmoceras* begin similarly, but the male and female become markedly different in the ultimate whorl. In this species, the male microconch shells have a long, narrow protrusion on either side of the aperture. However, given the backward water-jet propulsion of ammonoids, it's unlikely this adaptation increased their drag, so it wasn't just an ornamental adaptation, like those in a peacock. Just like in lappeted male perisphinctoids, this sexually selected trait may have been to help females differentiate males from juvenile and sexually immature members of their own species (Figure 7.18). The protrusion, called a lappet, only began to grow once the rest of the shell had stopped growing.

The first-appearance datum of the genus *Kepplerites* (Figure 7.19) formed the basis for the start of the Callovian, the last stage of the Middle Jurassic.

Figure 7.18 *Kosmoceras* **sexual dimorphism and sutures. Mature male with lappets on left.**

Figure 7.19 **Kepplerites.**

Lytoceratoidea

This group of ammonites represents a significant pulse in Jurassic ammonite evolution. Just like the phylloceratids and their lineage, the lytoceratid branch underwent a gradual evolution, but one that was incredibly prolific, from the beginning of the Jurassic and into the Cretaceous. A descendant taxon of psiloceratids, made up a large part of a lineage that eventually resulted in perisphinctids, stephanoceratids, spiroceratids, and eventually, the ancyloceratid heteromorphs.

The type genus, *Lytoceras*, is known for developing a highly evolute planispiral (Figure 7.20), with early forms whose divergences are barely detectable at the spiral level. However, the suture geometry of lytoceratids tells an important story. Up to now, the saddles and lobes of the phylloceratids were three-tined. In lytoceratids, the trident was replaced by a boomerang-shaped saddle and lobe.

The lytoceratids also were unusual among Jurassic ammonites in the complexity of their sutures, which were heavily frilled and averaged more than 20% greater complexity than their contemporary taxa. With this high complexity came larger and fewer saddles and lobes (Figure 7.21). Where many ammonites complexified by adding umbilical lobes, lytoceratids, never having more than a few umbilical lobes, just folded the ones they had.

Figure 7.20 (a) *Lytoceras taharoaense,* exhibiting whorl opening in the adult phase and (b) Mermaid for scale.

Figure 7.21 A sutured *Lytoceras.*

The exact reason for above-average complexity in ammonite sutures has been fiercely debated in the literature for generations. Ziegler placed lytoceratids, along with the phylloceratids, at the edge of the continental slope, living as deep as shelled cephalopods possibly can. This idea resurfaced when Gerd Westermann assigned lytoceratids the role of vertical migrators (Figure 7.22).

It was later shown by Robert Lemanis that highly complex sutures may not increase the structural integrity of a thin shell enough for very deep water. In ammonoids, it was argued, whose shells are much thinner than *Nautilus* shells, even shells with complex sutures were just as crushable as more simply sutured shells.

Figure 7.22   **A juvenile *Lytoceras* vertically swimming under a group of *Dactylioceras*.**

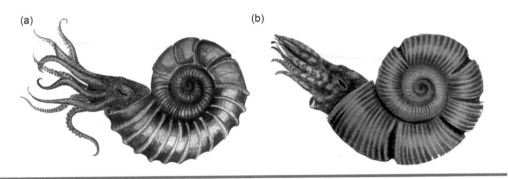

Figure 7.23 **(a)** *Alocolytoceras itzmeri* **reconstruction. (b)** *Alocolytoceras gr.*
germaini.

It has also been suggested that the complexity and relative thinness of ammonite shells was due to rapid secretion of the shell in order for ammonites to evolve tachytely, the speeding up of a species' lifespan. However, some of the most complex sutures of all time are associated with ammonoids whose growth bands indicate they had very long lives, in rare cases spanning centuries. Testing has shown that despite the frilliness of the suture, ammonoid shells did not *always* possess more tensile strength than nautilus shells. German researcher Helmut Keupp found that extremely complicated sutures can more easily support injured ammonites. If a shell was broken, an association probably existed, Keupp argued, between septal complexity and shell replacement speed. Models created by American researcher David Peterman have shown that increased complexity increases the capillary control over chamber fluids. There may have been an additional function to septal complexity. In 2012, Helmut Keupp showed a strong correlation between high complexity of sutures and apparent self-healing pathologies on ammonite shells.

Lytoceratidae and their subfamily, the alocolytoceratids, are characterized by a type of broad ribbing called "capricorns;" their exterior ornamentation resembles the texture of a goat's horn. *Alocolytoceras* were small, just a few centimeters in diameter, and simply sutured (Figures 7.23). *Lytoceras* shells could become over a meter in diameter, and even in small species or early in ontogeny, its sutures were exceptionally subdivided.

The End of the Jurassic Seapark

By the Late Jurassic, the first giant colonial cone-shaped rudist clams had appeared. The original Jurassic superfamilies of ammonites each had many thousands of generations of descendants, which had diversified and filled the oceans with extremely specialized new cephalopods. In contrast to many of the other cataclysms which claimed ammonoids previously, the end-Jurassic extinction is no longer considered a "mass extinction." The end-Jurassic event was particularly mild for ammonites relative to other extinctions which hit them much harder. Still, this "minor extinction" marked a shift in the life habits of animals, and ammonites were particularly influenced.

Enough ammonites and other animals made it through the end of the Jurassic that we now think of it as less an apocalypse and more a gradual change. The total number of genera was relatively constant through this transition, and actually, more new species evolved than disappeared. As with the end of the Triassic, ammonites overall were so abundant, and so ubiquitously saturated the paleoenvironment, that they provide a very high-resolution look at exactly what happened. For this reason, they are sometimes referred to as the "black boxes" of Mesozoic extinction events. The (S, D) (Figure 3.3) coordinates of surviving end-Jurassic ammonites were similar to those which survived the Permian, but perhaps more notably, they were wildly different from the values of end-Jurassic nautiloids. By this time in natural history, nautiloids and ammonites were so phylogenetically distanced that their habitats and modes of life would have been incredibly different.

Like other extinction events, volcanism and continental rifting were at the root of the cause of the Late Jurassic extinctions. However, it took decades to reach a worldwide agreement upon the Jurassic-Cretaceous boundary. Unsurprisingly, ammonites were relied upon to determine a boundary that geologists could agree on. Sowerby's original (1842–1851) boundary cited the last appearance datum for *Ammonites giganteus*. In the 1960s, a replacement boundary was proposed, at the base of units containing *Pseudosubplanites grandis* and *Berriasella jacobi*. In the 1990s, the first-appearance datum of *Dalmasiceras dalmasi* and *Riasanites* was apparently correlated to the *B. jacobi* boundary. The *B. jacobi* and *P. grandis* boundary is still generally seen as the closest approximation of the Jurassic-Cretaceous bounary.

Further Reading

Arkell, W.J. (1950). A Classification of the Jurassic Ammonites. *Journal of Paleontology*, 24(3), 354–364.

Davis, L.E. (2009). Mary Anning of Lyme Regis: 19th Century Pioneer in British Palaeontology. *Headwaters*, 6(14).

Hoffmann, R. (2010). New Insights on the Phylogeny of the Lytoceratoidea (Ammonitina) from He Septal Lobe and Its Functional Interpretation. *Revue de Paléobiology*, 29(1), 1–156.

Long, W.D. (1936). The Green Ammonite Beds of the Dorset Lias. *Quarterly Journal of the Geological Society*, 92, 423–437.

Maddra, R. (2015). Bitten Ammonites from the Upper Lias Group (Lower Jurassic) of Saltwick Bay, Whitby, North Yorkshire, UK. *Proceedings of the Yorkshire Geological Society*, 60(3), 153–156.

Maisch, M.W., Hoffmann, R. (2017). Lytoceratids (Cephalopoda, Ammonoidea) from the Lower Posidonienschiefer Formation (Tenuicostatum Zone, Early Jurassic) of Baden- Württemberg (South-Western Germany). *New Yearbook of Geology and Paleontology, Papers*, 283, 275–289.

Yacobucci, M. (2015). Macroevolution and Paleobiogeography of Jurassic-Cretaceous Ammonoids. In: Klug, C., Korn, D., De Baets, K., Kruta, I., Mapes, R. (eds) Ammonoid Paleobiology: From macroevolution to paleogeography. *Topics in Geobiology*, vol 44. Springer, Dordrecht.

Chapter 8

The Cretaceous Period: The Golden Autumn of Ammonites

"CHALK," a haiku by Alex Bartholomew:
Coccolithophores:
ground up and mutilated,
tortured into words.

Chalk

The Cretaceous is the longest period in geologic history; it spans 80 million years (from 66 to 146 Ma). More time passed from the beginning of the Cretaceous to its end than has passed from its end to now. It was also the period of ammonoids' greatest biodiversity and most extreme specialized traits. The Cretaceous derives its name from the Latin word *creta*, which means "chalk," since the Cretaceous is famous for its white chalk deposits. Chalk is a sedimentary rock made of the microscopic shells of coccolithophores (Figure 8.1), unicellular phytoplankton that produce a calcareous shell. Over thousands or millions of years, the microscopic shells accumulated and became compacted together. Because coccolithophores, like the ammonites and all mollusks, produce their shells from calcium carbonate, chalks are an extremely fine-grained limestones. Chalk deposits are often several hundred meters thick and can be found across Europe, especially in England, Belgium, Denmark, and France. The white Cliffs of Dover in England (Figure 8.1) are among the best-known chalk units.

DOI: 10.1201/9781003288299-8

Figure 8.1 **(a) The White Cliffs of Dover. (b) Coccolithophore fossils under microscope.**

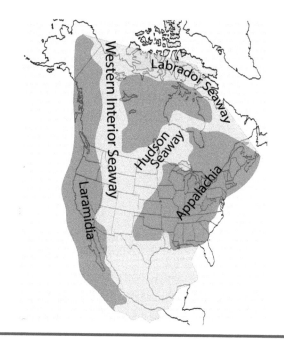

Figure 8.2 **Map of the Western Interior Seaway, which cut North America in two during the Cretaceous Period.**

In addition to the chalky seas of Europe, Cretaceous chalks are common in North America as well, from the Austin Chalk in central Texas to the Selma Chalk of Alabama and Mississippi. But the greatest expanse of chalk is in the Great Plains of North America, which were drowned by these same shallow seas through most of the Cretaceous. This formed a permanent marine barrier, the Western Interior Seaway (Figure 8.2), that lasted over 110 million years from the Middle Jurassic (about 180 Ma) until just before the end of the Cretaceous (about 70 Ma). It isolated the land and the dinosaur faunas of the world, so the dinosaurs of Montana have more in common with those from Mongolia than they do with the dinosaurs of New Jersey. The deposits of this great seaway can be found all over the western Plains and Rockies, from the chalk badlands of western Kansas to the Austin Chalk and other marine rocks of the Cretaceous of Texas and Oklahoma, to the thick deposits of marine Pierre Shale in South Dakota.

The chalk and all the shales and other marine rocks of the Cretaceous are a product of a much bigger phenomenon: the global greenhouse world of the Cretaceous. During the latter half of the Age of Dinosaurs, the climate was so warm that dinosaurs and crocodilians roamed above the Arctic and Antarctic circles. There was virtually no ice and snow anywhere at that time, since the atmospheric carbon dioxide levels were about 2,000 ppm (today it is over 415 ppm and climbing). The melting of all those icecaps resulted in extremely high sea level, and shallow epicontinental seas drowned many of the continents.

What could have caused such a warm greenhouse climate, and such an extensive flooding of the continents in the Cretaceous? There are several things happening at this time, all of which contributed to the rising sea levels and greenhouse atmosphere, although it is difficult to tell which were most important:

Breakup of Pangea: During the 80 million years of the Cretaceous, nearly all the remaining pieces of Pangea broke up and separated into the modern continents. The North Atlantic was already starting to open in the Late Jurassic, but most of its widening occurred in the Cretaceous. The opening of the South Atlantic—the separation of South America and Africa—occurred entirely in the Cretaceous, as did the separation of Africa from Antarctica, and the beginning of the split between Australia from Antarctica. Most impressively, India ripped away from its former position in Gondwana and raced across the Indian Ocean, so that by the end of the Cretaceous it was close to colliding with Asia.

This breakup of all the Pangea continents into their modern configurations had several important effects. The most important is that after rifting, the edges of the continents are wider and more stretched out, so their average continental elevations are lower, allowing the seas to flood them. These produce thick sedimentary wedge deposits on the edges of the rifting continents. For example, the bulk of the enormous passive margin wedge over 6,100 m (20,000 ft) thick on the Atlantic Coast and Gulf Coast of North America was formed during the 80 million years of the Cretaceous. These sinking continental edges were continuously drowned by shallow seas, producing enormous piles of marine shales and limestones that are today found all over the Atlantic and Gulf Coastal Plains. Much of the drowning of the continents is a product of the sinking of passive margin wedges.

Rapid Seafloor Spreading and Ridge Volume: When seafloor spreading is very rapid, it produces much thicker, taller profiles of rock on the mid-ocean ridge complex. This probably happened in the Cretaceous, when all the Pangea continents were pulling apart at a record rate of seafloor spreading. The increased volume of all those rapidly spreading ridges would have been enormous, and all that extra rock volume made the ocean shallower and pushed the water up onto the land. Numerous studies have shown extraordinarily high rates of spreading starting in the Early Cretaceous (Aptian Stage, about 126 Ma), and continuing at lesser rates until nearly the end of the Cretaceous.

How do we know that seafloor spreading occurs? The mid-ocean ridges run across the entire seafloor like the seams of a giant baseball, the longest chain of mountains in the world, covering over 65,000 around the globe. A massive fissure-like structure runs down the middle of the mid-ocean ridge, we see along the trench that new basalt (rock which forms the oceanic tectonic plates) is constantly being formed: magma from inside the earth is expelled here, extruding and producing brand new igneous rock. As new rock forms, older basalt is pushed away from the trench. Eventually, tectonics push the oldest, furthest away basalt back under the core, re-melting it and effectively repeating the cycle. This process has been going since the modern Atlantic opened in the Late Jurassic, and it is possible to use the magnetic signatures of these seafloor rocks to date them. By far, the vast majority of the rocks on the modern seafloor date back to the Cretaceous, thanks to both the high rate of spreading, and also the fact that the Cretaceous lasted 80 million years.

Greenhouse Gasses: Not only are high rates of seafloor spreading important in terms of the depth of the ocean, but the eruption of these volcanoes along the mid-ocean ridge released a lot more greenhouse gasses from the mantle. This was a major reason that atmospheric carbon dioxide levels may have reached values as high as 2,000 ppm (today it's only 415 ppm, even with all the greenhouse gasses we are introducing due to global warming).

Another contributor of greenhouse gasses was the indirect effect of oceans covering so much more land area than they do today. Rapid rates of tectonically induced mountain uplift caused rapid weathering, which is a major absorber of carbon dioxide. But with ocean water covering so much of the land surface, there were fewer areas of mountains rising around the world. The limited land area would have restricted the amount of weathering in soils, and thus how fast they could absorb carbon dioxide. If there had been any significant polar ice caps left in the Jurassic or Early Cretaceous, this extreme warming would have melted them all away, further contributing to the water in the oceans. (Those dinosaur movies that show glacial ice in the mountains in the background are wrong—there was probably little or no snow or ice anywhere in the world during most of the Cretaceous. Christopher Scotese, a paleogeographer, and his colleagues determined that annual mean temperatures throughout much of the Cretaceous were far too high to have snow just about anywhere on Earth.)

Mantle Superplume Eruptions: Recent research has shown that Cretaceous was a time of extraordinarily big eruptions of huge hot spots or mantle superplumes coming up from the lower mantle and then erupting beneath the oceans. These rocks formed by gigantic mantle eruptions were first discovered in the 1950s and 1960s, when oceanographers mapped huge submarine plateaus, especially in the Pacific Ocean. When scientists drilled and analyzed and dated the rocks, they proved to be the result of huge mantle lava eruptions that spilled enormous volumes of magma into the Cretaceous oceans—and pumped huge volumes of greenhouse gases from the mantle at the same time. The biggest of these is the Ontong-Java Plateau in the southwestern Pacific. About 126 Ma, it was formed by the eruption of 1.5 million cubic kilometers of lava (360,000 cubic miles) across the ocean floor that lasted less than a million years. This timing coincides with the huge spike in seafloor volcanism in the Aptian Stage of the late Early Cretaceous, and the rapid rise in sea level around the globe at that time. The Hess Plateaus and the Shatsky Rise also erupted during the Cretaceous, and their huge volumes of lava undoubtedly contributed to the Cretaceous greenhouse gasses as well.

Many factors may have contributed to the extraordinarily intense greenhouse world of the Cretaceous. Which was the most important is difficult to determine, but they were certainly all working at the same time, compounding their effects on global warming.

Cretaceous Reefs and Ammonites

Coccolithophores were not the only marine organisms which thrived during the Cretaceous Period. Carbonate platforms, similar to the shallow tropical waters around the modern Bahamas, thrived during these warm greenhouse

Figure 8.3 **(a, b) Cretaceous rudist reef. (c) Close-up of an individual rudist clam, shaped like a cone. (Courtesy M. Booth.)**

conditions, but they were different from the carbonate platforms that thrive in cooler climates. Relatively few stony corals littered the seascape, and strange cone-shaped clams known as rudists began to dominate the reefs (Figure 8.3) by accumulating and encrusting their skyscraping shells. Modern reefs are primarily built by coral, but reefs by definition do not need to include coral—they are defined as any biological structure produced by sedentary animals underwater that do not touch the surface, but come close enough to it that they affect the flow of surface water. Over Earth's history, reefs have been created by other organisms, including stromatolites, sponges, and bryozoans.

Ammonites are only rarely associated with coral reefs. The fragility of their shells was of as much concern for Cretaceous ammonites as it had been during the Jurassic, when their shells started to become especially thin. Being bashed against a coral or a rock was not a risk that ammonites could afford to take. However, as common continental slope dwellers (Figure 8.4), they can be associated with the steep slopes of the forereef (part of the reef which faces out to sea and is least affected by turbid water). Safely below the base of most waves, ammonites were protected from turbulent water. Even today the corals and sedentary creatures on the forereef are also the reef's most delicate creatures. This is the common home of the most delicate branching corals, which climb auspiciously into the water column at the forereef without the risk of near-constant destruction. The corals and mollusks living at the top of the reef, nearer the shore, are built for high-energy waves. Their skeletons are thick and massive, seldom branching in form or ornamentation.

There were a few ammonites that had a place near the shore, near the turbulent waters before the sloping reefs. The acanthoceratids dealt

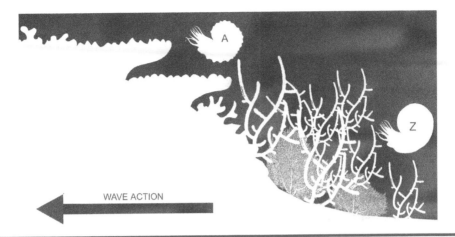

WAVE ACTION

Figure 8.4 **Cretaceous reef zones for ammonoids and corals. While corals were not as common reef constituents as rudists during the Cretaceous, they follow distinct patterns in reefs. The least sculptural are found in the hindreef, whereas the delicate branching varieties are found at the fore. In contrast, more ornamented Cretaceous ammonites such as *Acanthoceras* would be found nearer shore, and smoother shells like *Zelandites* would be found further out.**

Figure 8.5 **(a) Ornamentation in *Acanthoceras*, a relatively shallow-water taxon. (b) *Acanthoceras* swimming in inshore waters as pterosaurs fly overhead in the Western Interior Seaway.**

with their precarious position through the ornamentation of their shells (Figure 8.5). By the Cretaceous, many ammonites were smooth or only exhibited micro-ornamentation. Acanthoceratids borrowed traits from the past in order to gain entry to a new habitat. By re-evolving thick ribs and large nodes, acanthoceratids managed to increase their drag coefficient—despite floating freely, they would be less easy for violent waves to move.

Most acanthoceratid populations were far smaller than the more plainly ornamented ammonites of their time, including desmoceratids, gaudryceratids, and pachydiscids. This suggests that the acanthoceratid habitat was restricted, possibly due to their own specialization as the closest thing to a shallow-water ammonite. While smoother ammonites, such as gaudryceratids, are sometimes found in nearshore deposits, they are primarily found in assemblages that represent the offshore continental slope. Their presence in the near-shore waters was unlikely to have been a normal occurrence. Acanthoceratids were apparently so common in nearshore waters that they are frequently preserved among plant fossils on the periphery of the Western Interior Seaway. Either a minority of the gaudryceratids made it into the shallower water or their preservation there is purely taphonomic. Where they occur worldwide, acanthoceratids are almost always associated with facies nearer the shore than practically any other Cretaceous ammonite. Due to their apparent interfacing with paleoshorelines, they are sometimes the basis of biostratigraphic "biozones" in both the Western Interior Seaway of North America and the coast of England.

A Strange New World

We have a better understanding of the ecology of Cretaceous ammonites as living animals, and not simply fossils, than we do for Jurassic ammonites. Many genera and families of ammonites became globally distributed in the Cretaceous, but while Jurassic ammonites spanned full latitudinal intervals, Cretaceous ammonites also often branched out longitudinally. Pachydiscids, for example, can be found on the northeast coast of the United States, as well as Japan, but they are also known from Seymour Island at the tip of the Antarctic Peninsula. While the same Jurassic *Dactylioceras* may be found in England as is found at similar latitude in eastern Russia, you would be hard pressed to find it in Botswana. Cretaceous ammonites are commonly found all over the world, with the more common ones occurring on every continent. The worldwide abundance of these ammonites—scaphitids, baculitids, and too many others to name—has made them excellent index fossils.

One of the most common ammonites in the fossil trade is an genus called *Cleoniceras* (Figure 8.6). These ammonites are found in Albian (Early Cretaceous: about 113 to 100.5 Ma) deposits across the world, but especially

Figure 8.6 **Calcified *Cleoniceras* cross section.**

northern Africa and Europe. There are often so many shells of these ammonites and other mollusks (including the occasional *Eutrephoceras*, an early modern nautilus from the Cretaceous) preserved together that over millions of years, groundwater dissolves some of them and redeposits calcite crystals inside of other *Cleoniceras* shells.

The Atlas Mountains region of Morocco is a well-known trove of Early Cretaceous ammonites, such that fossil preparation is a major industry and source of jobs in the region. Agadir is one of the world's oldest localities from which countless heteromorphs can be found. Commonly gyrocones and ancylocones, these early heteromorphs have still barely deviated from planispirals, the "regular ammonite" shape.

Stephen Jay Gould coined the term "left wall of minimal complexity," from which all life starts out as relatively simple. Beyond the left wall, conditions were too simple for life to exist. Instead, life accrues both complexity and variation as they evolve and diversify and move progressively to the right, which represents the progression of time and typically increasing size and complexity as well. Ammonites showed this pattern throughout their long history, but they amped it up in the Cretaceous. There were almost as many ammonoid superfamilies in the Cretaceous as there had been at the peak of superfamilies in the Late Jurassic, but far more morphotypes. The descendants of the original Jurassic superfamilies had branched into unprecedented diversity. Some Jurassic genera, including *Lytoceras* and *Phylloceras*, were still alive. But many other families of ammonites were born from the Jurassic stock that would also become important index fossils. Biologically, they present their own

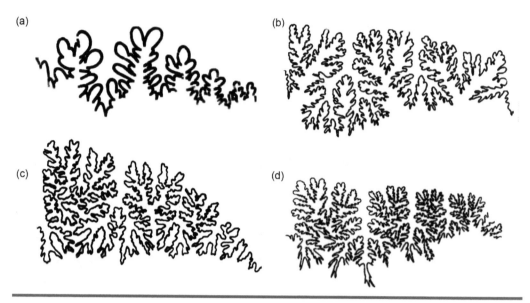

Figure 8.7   Suture patterns of ammonites that appeared from the Triassic to the Cretaceous (a) *Phylloceras*, in the Jurassic and Cretaceous (b) *Lytoceras*, in the Jurassic and in the Cretaceous (c) *Gaudryceras*, a distant Cretaceous lytoceratid descendant and (d) *Pachydiscus*, a distant Cretaceous desmocerid descendant.

challenges and puzzles. As far removed as many of the Cretaceous families were taxonomically, they sometimes exhibited convergent evolution. Each of these Cretaceous families that greatly complexified their sutures (Figure 8.7).

The Desmoceratids

Another descendant lineage of phylloceratids, the desmoceratids, appeared suddenly in the Early Cretaceous. While this family did not as quickly increase in septal complexity, they did increase diversity by expanding into a niche previously untapped by ammonoids. The desmoceratids eventually became a fairly diverse group of ammonoids, being found on almost every continent and growing to monstrous sizes.

The puzosiids—ammonites with shells of gigantic diameters, often longer than an adult human—are desmoceratids. Juvenile puzosiids were unassuming in size but may have possessed radial lirae. Adult *Puzosia mayoriana* were more than a meter in shell diameter, and despite their great size, the ancestral traits of pachydiscids are still present in their sutures (Figure 8.8a). *Parapuzosia bradyi* (Figure 8.8) had a shell diameter almost as tall as an average adult human. However, the macroconch

Figure 8.8 (a) A mostly complete specimen of *Puzosia mayoriana*. (b) Elasmosaurs hunt fish alongside giant desmocerid descendants. Size comparison of an adult mer-person with reconstructions of living *Parapuzosia*. (c) Zachary Prothero with a cast of the type specimen of *Parapuzosia seppenradensis*. (d) Dozens of puzosiids and giant ammonites in the Mikasa City Museum, Hokkaido, Japan. (a from K. Marriott; c from D. Prothero; d courtesy Wikimedia Commons.)

of *Parapuzosia seppenradensis* (Figure 8.8) was the largest ammonite species of all time, measuring at least 175 cm (6 ft) in diameter. Given that we have no idea how long its arms were, we can only assume puzosiids were much more massive creatures than even their shells suggest.

Desmoceras (Figure 8.9) had a similar coil and expansion rate to *Pachydiscus* but often, much simpler suture patterns. The type genus of the desmoceratids appeared a few million years after the puzosiids first

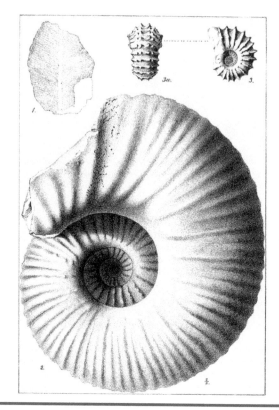

Figure 8.9 *Desmoceras,* courtesy Geological Survey of Canada, from J.F. Whiteaves (1900).

arose, and unlike puzosiids, which endured to the end of the Cretaceous, it did not stand the test of time. For this reason, they are considered a great index fossil for the uppermost Albian stage of the Cretaceous. Their immediate role in the Lower Cretaceous is at the Albian-Cenomanian Boundary—an extinction event that was less severe than true mass extinctions, but significant enough to delineate a new stage in the Cretaceous period.

Minor Extinctions (Oceanic Anoxic Events) in the Cretaceous: The Albian-Cenomanian Boundary

The extinction event at the end of the Cretaceous Period is probably the best-known mass extinction of all time. Having claimed the non-avian dinosaurs and ultimately giving way to large mammals, the K/Pg event is the only mass extinction to have become widely known in popular culture despite the fact that it was relatively small compared to other mass extinctions, such as the end-Permian event. The K/Pg extinction only accounted for about a quarter of all biodiversity that was lost in the Cretaceous. Throughout the long saga of the Cretaceous, numerous "minor extinctions" occurred, dividing time and giving us a basis for the various stages of the period. The "minor" in "minor extinction" is a bit of a misnomer. Minor extinctions are large-scale extinction events which claim multiple taxa but are generally not as devastating to biodiversity as mass extinction events. The minor extinctions of the Cretaceous were often tied to climate and are evidenced by sudden rises and falls in sea level. They were especially felt in marine ecosystems. Many of the events which triggered significant extinction throughout the Cretaceous were called oceanic anoxic events, or OAEs. The OAEs are numbered in the order of the sequence in which they happened.

The earliest of these, called OAE 1a, began in the Aptian Stage. The next part of the first oceanic anoxic event, OAE 1b, occurred at the Albian-Cenomanian Boundary, about 101.5 million years ago. OAE 1b significantly impacted ammonites. The Albian is considered the last stage of the Early Cretaceous. (Despite the long length of the Cretaceous Period, there is no formal "Middle Cretaceous" epoch.) On the standard geological time scale, it is just split into Early Cretaceous and Late Cretaceous at almost exactly the midpoint. The Albian-Cenomanian Boundary is the point at which this happens. Its exact stratigraphic placement was determined using ammonites as index fossils. The end of the Albian was marked by rapid global cooling. Its biotic shifts are correlated to global regression, meaning that sea level dropped, and more sea water was stored in the ice caps. On top of this anoxia swept the oceans. The sudden drop in temperatures and loss of oxygen caused many organisms to die out locally, or even become extinct globally.

Cenomanian-Turonian Extinction

Later in the Cretaceous, another significant extinction event caused mass turnover of countless species of mollusks, including ammonites. OAE 2 marked the end of the Cenomanian stage, like most Mesozoic extinctions, and it represents a period of global warming. Supervolcanism in the Caribbean was pouring catastrophic amounts of greenhouse gasses into the atmosphere. The Western Interior Seaway became enormous, at some points covering more than half of North America's area. The seaway was rapidly expanding its areal coverage and en route to its maximum, which it reached in the Turonian stage. It is believed that the success of predators in the Early Cretaceous, such as marine reptiles and large fish, contributed to a gradual decline of ammonoids and other mollusks throughout the Cretaceous. Marine reptiles, which had thrived throughout the earlier Mesozoic, had established themselves in the seaway, exerting new evolutionary pressures on the mollusks.

The carbon dioxide and methane emitted from volcanoes and seafloor spreading centers rapidly melted away any ice there was in the world, so sea level rose and became warmer. The event was sharply punctuated and represents one of the most extreme "intra-Cretaceous" minor extinctions. Ammonites and inoceramid clams were shocked by this sudden change in temperature and depth, and most were unable to keep up with it. Still, a more deadly threat loomed for the mollusks that managed to survive. Warming oceans are much saltier than cooler water. Because warm water molecules have more space between them, they have greater capacity for salt molecules. As a result, salinity spiked at the Cenomanian-Turonian Boundary.

European marine faunas above and before the boundary are distinct, sharply beginning and ending right at this time—they consist mainly of ammonites, belemnites, various fish, and rudist bivalves. (This was actually what enabled d'Orbigny to delineate and name the stages of the Cretaceous.) Not surprisingly for an event characterized by a shoreline migration, it is an acanthoceratid ammonite, *Watinoceras* (Figure 8.10), which marks the boundary in the Western Interior Seaway of North America. *Watinoceras* appear abruptly above the boundary in the earliest Turonian stage.

Watinoceras is also part of the early Turonian in the Western Interior Seaway of North America. As the Western Interior Seaway covered its greatest area during the Turonian, at the time of the Cenomanian-Turonian extinction, it was running all the way down North America. The boundary's sharp contrast is

Figure 8.10   (a) *Watinoceras.* (b) *Watinoceras* reconstruction.

especially palpable in the Western Interior Seaway, whose calm waters provided havens, and later, protected entombment, for many of its creatures, ranging from the tiniest coccolithophores and foraminifera to the largest marine reptiles. Here, ammonites still provide a pivotal glimpse into the dramatic turnover, but they are well supported by the rapid extinction and replacement of not only inoceramids, but foraminifera, decapod crustaceans, and large swimming vertebrates opening to the Tethys Ocean again in Mexico. This connection resulted in an unprecedented amount of faunal diversity in the Western Interior Seaway. The boundary is locally defined primarily by the first appearance of the inoceramid *Mytiloides puebloensis*. Ammonites which helped herald the early Turonian at the Vallecillo Lagerstätte of northeastern Mexico include *Pseudaspidoceras flexuosum*, *Watinoceras coloradoense*, *Vascoceras birchbyi*, and *Mammites nodosoides*. The first appearance of several sharks, which are almost all known from their teeth, numerous teleost fish, and several new Cretaceous sea turtles lend further reinforcement to this recovery after an extinction event.

Turonian-Coniacian Extinction

One more particularly abrupt turnover of ammonites would occur before the Cretaceous was over. Around the world, from Russia to Antarctica, stark last-appearance events at the end of the Turonian and sudden first appearances of the Coniacian fauna are found in the ammonite and inoceramid assemblages. On the northern Japanese island of Sakhalin, the Turonian faunas completely disappear and are quickly replaced. *Inoceramus multiformis* and the relatively common ammonites *Jimboiceras planulatifonne* and *Scaphites planus* disappear. Interestingly, the most apparently specialized ammonite in this fauna—the famous heteromorph, knotted *Nipponites mirabilis*—is one of a select few species that make it through to the Coniacian. Though the most likely cause of this extinction was global warming, *Nipponites* was probably a planktonic drifter. It made it through, despite inhabiting waters that would certainly have most heavily felt the brunt of the climate disaster.

As with the previous intra-Cretaceous "minor" extinction, the opening of the Western Interior Seaway to the south enabled new fauna to move in from the open ocean. As a result, the inoceramid assemblages at the Turonian-Coniacian boundary in Mexico and the United States are virtually identical to those in Europe. In the Western Interior Seaway, this penultimate turnover for ammonoids is represented best by the Pueblo and La Junta regions of Colorado, as well as localities in northeastern New Mexico.

Evidence for the Turonian-Coniacian boundary is exceptional at the Múzquiz Lagerstätte, also in northeastern Mexico. The impeccable fine-detail preservation of vertebrates has been compared to the famous Upper Jurassic Solnhofen Limestone of Germany, which produced the first specimens of pterosaurs and all known specimens of *Archaeopteryx*, among other incredible fossils. Foraminifera of the Múzquiz Lagerstätte are integral to the biozones of the Turonian and Coniacian even though they are considered rare, and often just a single specimen represents them. Due to their abundance, the inoceramids *Mytiloides scupini*, for the late Turonian, and *Cremnoceramus deformis erectus*, for the early Coniacian, represent the fallout. Though ammonites exist here, they become rare. Ammonites and inoceramids were used to determine the end of the Coniacian and beginning of the next stage of the Cretaceous, the Santonian. In the Western Interior Seaway, scaphitids became the most dominant members of the ammonite fauna through what we call OAE 3. Scaphitids were probably strong swimmers compared to other ammonites, and could more easily migrate to an area of seaway that was more oxygenated than the ammonite taxa that were lost at this interval.

The Late Cretaceous

Placenticeras

Ammonites in the genus *Placenticeras* had platycone (laterally compressed almost to the point of complete flatness) shells. Only mature females, which probably required the space for egg brood pouches, developed any roundness in the part of the shell surrounding the body chamber (Figure 8.11).

We used to believe that *Placenticeras* were some of the fastest swimmers of the ammonite world due to their lateral compression. Their flat shells appeared able to have slice through the water like knives (remember, ammonoids swam backward) and their tight shell coils would have produced almost no drag. However, a team from the University of Utah showed that the flat shape of *Placenticeras'* aperture and therefore, its soft body, would have probably been unable to produce enough thrust for the fast-swimming model of flat-shelled ammonites (oxycones and platycones) to be true.

Using Raup's coiling system (Fig. 3.3), we can see that *Placenticeras* had very high W and very low D; in essence, their whorl section grew more rapidly versus the amount of new shell they were adding. This contributed to the flatness of their bodies. It also results in a highly involute shell, with the last whorl covering most of the whorls beneath it.

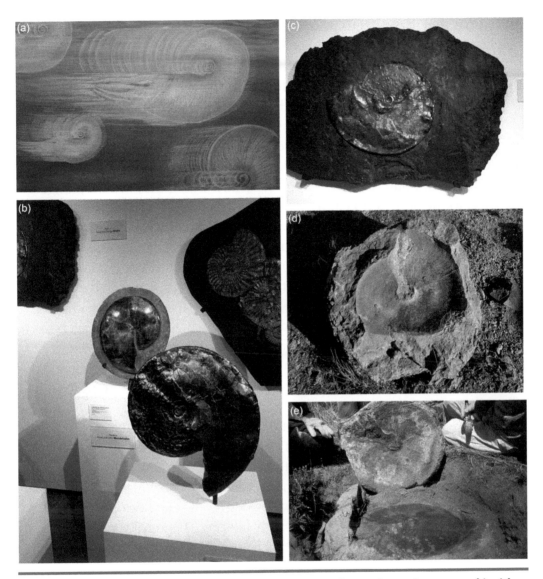

Figure 8.11 (a) *Placenticeras intercalare* ammonites. (b, c) *Placenticeras meeki* with ammolite from the Bear Paw Formation, Alberta, Canada. (d-e) Huge *Placenticeras* macroconchs weathering out of the Pierre Shale north of Kremmling, Colorado. (d and e Courtesy E. Evanoff.)

One of the most unusual collecting areas for *Placenticeras* is in the Middle Park Basin north of Kremmling, Colorado. Across a sweeping landscape of sagebrush, there lies one of the richest fossil beds in the Rocky Mountains. As you walk across the flats, you come upon one gigantic ammonite shell after another, some almost a meter across (Figure 8.11d and e). They are weathering out of the underlying Pierre Shale, and many are just lying on the surface, or just slightly buried. They are so huge, and their casts in sediment form such a distinctive concave top surface, that they have been nicknamed "bird baths." In addition, there are also gigantic inoceramid clams that once lay on the same sea bottom, and scaphitids and baculitids are among the most common fossils. When a large crew from the Denver Museum of Nature and Science and the University of Northern Colorado was collecting there, they realized that there must be many more of these huge shells lying just beneath the surface. They rigged up a probing device made of sharpened metal rods (like a giant pitchfork) and systematically probed their way across the thin soils formed from the Cretaceous shales. Sure enough, every once in a while, they would plunge the rods into the soil, and hear a "clunk," and dig up a beautiful specimen that had not yet reached the surface or begun weathering.

The huge number of large shells suggests that most of the fossils are macroconchs (the larger female shells) of *Placenticeras*. There are about 13 females to every male in the site. It is thought that they may have been congregating in huge numbers on the Cretaceous seafloor, possibly brooding their embryos, when they were buried and fossilized. The site has been studied intensely by Dr. Emmett Evanoff and his students at the University of Northern Colorado. As Evanoff put it, "That 13 to 1, female-to-male ratio suggests this was not a catastrophic event," he said.

> When we were digging up some of the ammonites, the lower sides are complete while the upper sides of the shells have been chipped away by marine scavengers. What this suggests to me is that this was a brooding ground where the ammonites, who probably mated only once in their lives and then died like modern squids and octopus, came here to lay their eggs in the sandy bottom. The females probably guarded their egg clutches until they fell over on their sides and died.[1]

Placenticeras are also one of the most common producers of a rare semi-precious gemstone called ammolite (Figure 8.11b-c). The iridescent rainbow effect of ammolite is a product of differential weathering on the surface of the shell's nacreous "mother-of-pearl" layer made of the mineral aragonite, which has been exposed to precipitation over millions of years. Different tablets (each about just 0.5 microns thick) of the nacre reflect different wavelengths (parts) of the color spectrum, similarly to an oil puddle in a parking lot.

---

[1] https://www.skyhinews.com/news/72-million-years-ago-kremmling-cretaceous-ammonite-locality-takes-trekkers-into-the-past/

Other ammonites, commonly baculitids and scaphitids, can form ammolite, but generally are only considered true ammolite if they come from the Bear Paw Formation in Canada. The gemstone's longest running cultural prevalence is as *Iniskim*, or "buffalo stones," talismans of good luck for hunters from the Blackfoot Nation. The 2018 documentary film *Ammolite: Gem of the West* features expert insights from Troy Knowlton, a lifelong collector of ammolite and member of the Blackfoot Nation, as well as paleontologists from Canada's Royal Tyrell Museum.

The stone has similar cultural significance to the Kainai (Blood) Tribe, who have worked hard to protect the quarries in which these special ammonites are found from vandalism and over-harvesting of the fossils. To collect ammolite in the region, one must obtain a Surface Mining Collection Permit from the Blood Tribe reserve 148.

Today, a complete *Placenticeras meeki* (easily 60 cm in diameter) may retail at a quarter million US dollars. Even small pebbles of ammolite can be worth hundreds of dollars. Apart from the size and the completeness of the fossil, color is also a factor in determining the value of ammolite. Reds and greens are the most common colors, but specimens that have large patches of blue and purple are the most valuable.

Ammolite is similar to opal in terms of its relative softness. To become suitable for use in jewelry, it needs to be stabilized, or else it will easily crack. Nearly all specimens of ammolite, for jewelry or display purposes, have been stabilized with invisible clear coatings that prevent them from breaking.

Wes EagleChild is an ammolite surface miner who has developed incredible skill discovering ammonites bearing the gemstone in the cliff faces of the Bear Paw Shale. EagleChild embarks on his quests early in the morning and regularly uncovers massive *Placenticeras* ammonites, including those in the coveted "blue zone." Traversing vertical drops over 100 m tall in all weather conditions, he expertly protects the shells from destabilizing while he is still uncovering them.

Baculitidae

Worldwide, baculitids are one of the most abundant heteromorph families. Their simple, straight shells and rapid life cycles ensured the success of baculitids throughout the Cretaceous Period. After the Devonian Period, ammonoids became what are called r-selected species, which are organisms that have lots of offspring which are not nurtured by their parents against predators, but instead their numbers swamp the predators that feed on them. While all ammonites could be considered r-selected, baculitids took this to an extreme. The suggestion is that baculitids were low-ranking and fast-reproducing secondary consumers on which many marine paleoenvironments relied heavily. Their sheer abundance and rapid speciation have made them excellent index fossils, such that they are commonly used to delineate biostratigraphic zones. The *"Baculites* biozones" of the Pierre Shale are one such example (Figure 8.12). Less prolific ammonite genera such as *Didymoceras* tend to mirror the extinction and speciation events of baculitids in these successions.

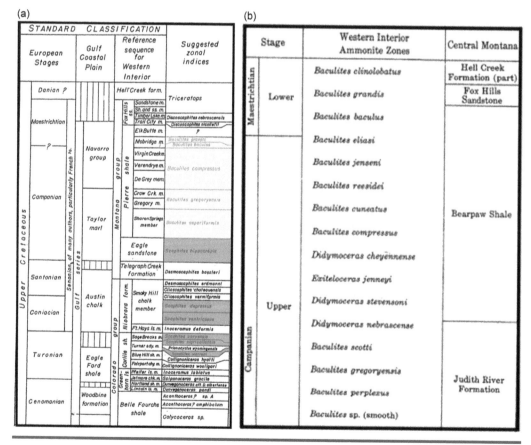

Figure 8.12 (a) *Baculites,* along with *Scaphites,* make up the biozones of the Western Interior Seaway and show rapid extinction and replacement cycles–the Campanian, for example, is only about nine million years. (b) Baculites species as index fossils in the *Baculites* biozones.

Sciponoceras

Inevitably, the soft-tissue of ammonoids has been a very challenging issue for researchers because it very infrequently preserves. The best currently known examples of ammonoid soft tissue preservation are from ubiquitous baculitids. We know from their fossils, for example, that baculitids probably had eyes more like those of coleoids than those of nautiloids or gastropods. A 2012 discovery including multiple *Sciponoceras* (Figure 2.6) with exceptional soft-tissue preservation documented by Christian Klug and colleagues, confirmed once and for all that baculitids had arms, though from this discovery alone, the exact lengths of the arms were indeterminable.

Baculites

As adults, *Baculites* ranged in size from a few centimeters to several meters (Figure 8.13). They primarily lived with their heads pointed down, vertically suspended in the water column, though their preferred water depth may have differed based on species and ontogenetic stage. Based on their stable-isotope ratios, we know that different genera of baculitids lived both pelagically and benthically. Some may have not had strong preferences for water depth. Occasional benthic gastropods were found in the stomach contents of baculitids, whose prey mostly consisted of planktic crustaceans. Whatever their preferred depth, baculitids likely all had similar life habits. Baculitid radulae have microscopic crustaceans preserved as they were caught in radular teeth of *Baculites*, so they are known to have filter-fed zooplankton. Their upper beak was much smaller than their lower jaw, suggesting that they filter fed continuously throughout the day. Like most heteromorphs, baculitids had smooth, wide beaks that featured a spout-like central invagination. The radula of a *Baculites* was reconstructed by Kruta and colleagues in 2015, using high-resolution computed tomography, a technology which paleontologists have occasionally borrowed from the medical field. The individual radular teeth of *Baculites* are longer and more needle-like than the radular teeth of modern coleoids, who are usually excellent hunters of other marine macrofauna, or of gastropods (Figure 2.5).

Scaphitidae

The scaphitids appear superficially similar to planispiral ammonites (Figure 8.14), or to very compact ancylocones. However, they are much more

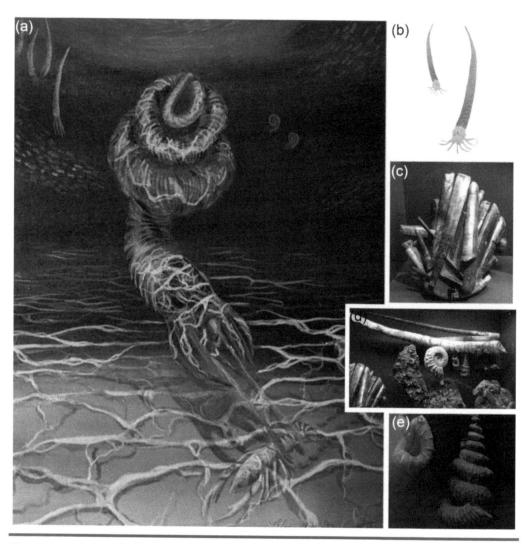

Figure 8.13 **(a) Baculites drifting above a benthic nostoceratid. (b) Reconstruction in life orientation. (c) A cluster of *Baculites* shells preserved roughly parallel to one another after death. (d) Nearly complete long tusk-like shells of *Baculites*, plus other Cretaceous ammonites. (d) Close-up of nostoceratids from the Denver Museum.**

derived. Like baculitids, they are remarkably abundant in numerous marine deposits from the Upper Cretaceous. Their simplicity relative to other hetero-morphs may have been a cause for their enormous evolutionary success. As genera and species were short lived, they have frequently been used as index fossils. Different scaphitid genera are commonly differentiated by counting the rows of nodes lining the shell. *Hoploscaphites* commonly has three rows of nodes, for example, whereas *Discoscaphites* has five. In the Western Interior Seaway, *Hoploscaphites* were especially abundant in the benthic oases around methane seeps (Figure 8.14), rare areas of seafloor where oxygen was plentiful.

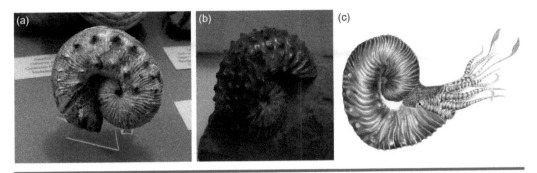

Figure 8.14 **(a)** *Hoploscaphites nodosus,* **(b)** *Hoploscaphites nicolletti,* **and (c)** **Reconstruction of *Hoploscaphites nodosus.***

Hoploscaphites

Hoploscaphites (which, along with *Discoscaphites* used to be called *Jeletzkytes*) are a common occupant of the deepwater facies of the Pierre Shale. Stable-isotope ratios based on benthic and pelagic foraminifera have shown that *Hoploscaphites* began its life in the benthic ecosystems and flourished around methane seeps, but then migrated into shallower waters as it became a juvenile. As adults, they periodically returned to the benthos to breed. All of this is recorded by the chemical isotopes in the growing chambers of their shells.

As early as the middle of the Jurassic, another group of ammonoids had revolutionized ammonoid paleoecology. Though they were a main staple of both Jurassic and Cretaceous marine fossil assemblages, we will discuss these strange ammonites independently in the following chapter.

Further Reading

Ammolite: Gem of the West. 2019 Film.
Black Horse, F. (2019). Blood Reserve Rockhunter Brings Iridescent Fossil Shells 'Back to the World'. *CBC News.*
Cobban, W.A. (1984). Mid-Cretaceous Ammonite Zones, Western Interior, United States. *Bulletin of the Geological Society of Denmark*, 33, 71–89.
Everhart, M.J. (2005). *Oceans of Kansas: A Natural History of the Western Interior Sea.* Indiana University Press, Bloomington, IN.
Hoffmann, R., Slattery, J.S., Kruta, I., et al. (2021). Recent Advances in Heteromorph Ammonoid Palaeobiology. *Biological Reviews*, 96(2), 576–610. doi: 10.1111/brv.12669.

Klug, C., Riegraf, W., Lehmann, J. (2012). Soft-Part Preservation in Heteromorph Ammonites from the Cenomanian-Turonian Boundary Event (OAE 2) in North-West Germany. *Palaeontology*, 55(6), 1307–1331. doi: 10.1111/j.1475-4983.2012.01196.x.

Landman, N.H., Cobban, W.A., Larson, N.L. (2012). Mode of Life and Habitat of Scaphitid Ammonites. *Geobios*, 45(1), 87–98.

Larson, N., Jorgensen, S., Farrar, R.A., Larson, P.L. (1997). Ammonites and the Other Cephalopods of the Pierre Seaway: Identification Guide. Geosciences Pr.

MacLeod, N. (2015). *The Great Extinctions: What Causes Them and How They Shape Life*. Firefly Books, London.

Skelton, P.W., Spicer, R.A. (2003). *The Cretaceous World*. Cambridge University Press, Cambridge, UK.

Wright, C.W. (1953). XLV—Notes on Cretaceous Ammonites. I. Scaphitidae. *Annals and Magazine of Natural History*, 6(66), 473–476. doi: 10.1080/00222935308654446.

Chapter 9

Ancyloceratoidea: The Heteromorphs

Heteromorphic ammonoids are the rare deviants whose shells coiled differently from the long history of "planispirals" we just considered. They are not widely known outside the community of paleontologists and fossil collectors, but their biology continues to provide a deep, largely undisturbed well of both scientific and speculative intrigue. It is not completely understood why these bizarre forms evolved, and the prevailing theories of how they lived have changed dramatically over past decades. Theories have ranged from the hardly creative to the absurd and unfounded. The least risk was taken by interpreters of heteromorphs in the 1960s, who envisioned them as snail-like benthic crawlers. This interpretation has sometimes invoked comparisons to living vermetid snails, irregularly coiling marine gastropods who are unable to swim and are so hydrodynamically cumbersome that they can barely glide along solid paths. Today, this interpretation can most commonly be seen in old dioramas at natural history museums. (It is important, when making comparisons between mollusks, to understand that snail shells are not chambered.)

When describing ammonites, the word "heteromorph" has two meanings. Commonly defined as simply, "differently shaped," a more accurate interpretation might be "many morphs," as each growth stage in the life of many species manifests as a different shape, or morph, being added to the shell. Most paleobiologists use the "many morphs" interpretation, because with each new revolution, the animal likely co-opted the next age-segregated niche in the ontogeny of its kind. Along with mobility, behavior, and niche partitioning, a

DOI: 10.1201/9781003288299-9

major constituent of the nature of heteromorphs is their ever-migrating center of gravity: the orientation of the shell itself changed over the animal's life cycle. Furthermore, the immense biodiversity preserved in their calcitic body parts may reflect unimaginable disparity in their arms, eyes, and even colors.

Heteromorphs first appear in the fossil record of the Late Jurassic, approximately 199 Ma, but they remained uncommon, and relatively conservative in shape until the Cretaceous.

Heteromorphs are broadly considered polyphyletic, in that heteromorph-like species independently evolved in multiple groups, but all true heteromorphs are members of the suborder Ancyloceratoidea. Some primitive forms include *Tropaeum* or *Australiceras* (Figure 9.1a and b). Over their evolution, heteromorph coils gradually opened more, and distinct phases of growth, called morphs, began to appear. Closed or fully open coils, straight shafts, and 180-degree hooks could all appear at different life stages in one shell, as in the Moroccan heteromorph *Ancyloceras vandenheckii* (Figure 9.1c)

(a)   (b)   (c)   (d)

]2.5 cm

Figure 9.1   (a) *Australiceras jackii,* (b) *Australiceras irregolare,* (c) *Ancyloceras vandenheckii,* and (d) *Rossalites imlayi.*

and the Italian *Rossalites imlayi* (Figure 9.1d). As the Jurassic ended and the Cretaceous began, heteromorph families started to expand into myriad forms that can be divided into at least 50 geometric categories.

Most paleobiologists now agree that the bizarre shells sported by heteromorphs (Figure 9.2) appear to simply be an example of niche partitioning taken to the extreme. Living cephalopods exhibit much more disparity in their soft tissue than hard parts. Likewise, the range of possible soft-part biodisparity may have been extreme. At the species level, high degrees of sexual dimorphism appear rampant, and with the different life stages of some

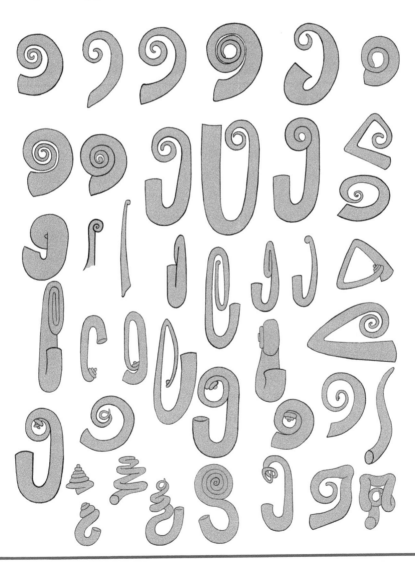

Figure 9.2 **Heteromorph ammonites account for nearly all of the roughly 50 shell shapes of ammonoids, based on Kakabadze (2015).**

Figure 9.3 **A shoal of krill floats through a bloom of *Nipponites mirabilis*.**

heteromorph shells varying drastically, it is plausible that soft-part morphologies and behaviors were equally as disparate. In the nineteenth century, social Darwinist Herbert Spencer, who famously coined the phrase "survival of the fittest," appropriated the seemingly aberrant diversity of heteromorphs in an attempt to bolster public support for his opinions. He likened the accelerating proliferation of weird shapes among the ammonoids leading up to the Cretaceous mass extinction event (the same event which wiped out the non-avian dinosaurs) to Nero playing his fiddle during the fall of Rome. Like Nero, Spencer argued, heteromorphs were not simply specialized, but an indication that ammonites had been long degenerating. Their waning fitness, then, brought about their total elimination during that extinction event. The inability to survive the end-Cretaceous extinction event by heteromorphs, and by extension, all ammonoids, was viewed as the product of their own inherent defects, and not a cataclysmic event that may have been set into irreversible motion long before the first ammonoid eggs even hatched over a quarter billion years before. We now know that most of Herbert Spencer's views were incorrect, including his speculation on the ammonite extinction based on how bizarre some of them became (Figure 9.3).

Paleoecology

The interactions of ammonoids with the paleoenvironments in which they are found can be determined using a combination of several methods.

Beyond basic shape, heteromorph ammonite fossils vary dramatically in their chemical compositions, faunal assemblages, and the types of sedimentary rocks in which they are now found.

When determining the life habit of an ammonoid, it is important to determine the living creature's position in the water column relative to other fossils alongside which it was found. Initially, all fossils in a bed are found together, and any original differences in depth have been lost and compacted by the process of sedimentary deposition. The propensity an ammonoid had for floating can be measured both physically and chemically. Physically, the size of the living chamber—and therefore, the size of the heavy portion of the ammonoid—relative to the size of the gas-filled phragmocone, can tell us whether an ammonite had positive, negative, or neutral buoyancy. Three-dimensional models with strategically placed weights can be used to determine the buoyancy of a living ammonoid. However, the depth at which it lived is usually obtained using stable-isotope ratios preserved in shells. Luckily, this information is seldom lost, as long as the original calcium carbonate of the shell is still intact. A common index fossil found alongside ammonites is the giant, flat Cretaceous clam *Inoceramus*, which lived on sea bottoms all over the world. Based on its size and shape, we can deduce that *Inoceramus* did not swim and was benthic. Therefore, we can compare the stable-isotope ratios of carbon-13 and oxygen-18 preserved in inoceramid shells to other shells, including ammonites. As the heavier of the common oxygen isotopes, oxygen-18 is more frequently found in colder water, including bottom water, because the lighter oxygen-16 is more easily evaporated, leaving behind a higher concentration of oxygen-18. Similarly, organic marine waste—which is high in carbon-13—eventually sinks to the bottom of the ocean rather than floating indefinitely. Therefore, shells that lived near the sea bottom are low in oxygen-18 and high in carbon-13. Shells that live closer to the surface will be increasingly high in oxygen-18 and low in carbon-13.

Figure 9.4 shows the placement of several heteromorphs endemic to the United States Pierre Shale based on stable-isotope analysis from inoceramids. Some ammonites favored different positions in the water column during different life stages.

Heteromorph fossils are not particularly common but can be locally abundant in restricted areas, or restricted genera. In contrast to planispiral ammonites of similar age, whose distributions can sprawl across continents, individual ancyloceratoid taxa normally come from just one or two rock formations. Sometimes, heteromorph-rich rock units house several unusual ammonite species. There are several primary locations for this rich heteromorph

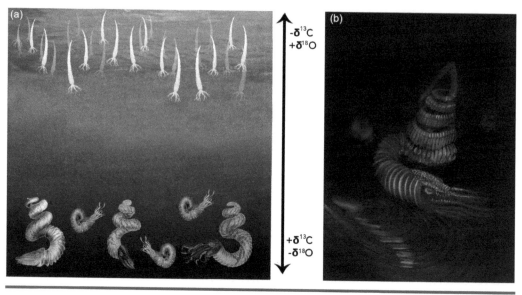

Figure 9.4 **(a) WIS ammonites in the water column based on stable-isotope ratios. Shells with isotopic ratios similar to *Inoceramus* are interpreted as epibenthic and shells with oxygen and carbon isotopes that are opposite the inoceramids are considered pelagic. (b) *Didymoceras*, a demersal genus, hovers above an inoceramid.**

biodiversity: the Atlas Mountains of Morocco, the Western Interior of North America, and two locations in Japan, the Yezo Group and the Izumi Group.

The Pierre Shale

This unit in the western United States encompasses the deepest lithofacies of an epicontinental sea that drowned much of Cretaceous North America, commonly called the Western Interior Seaway (WIS) (Figure 8.2). The units above and below the Pierre Shale—namely, the Niobrara Formation and the Fox Hills Formation—are sandstone, showing an oscillation of shallow into deep-water conditions that ultimately got shallower and shallower until the disappearance of the seaway sometime after the end-Cretaceous extinction.

The WIS is responsible for many of the world's heteromorphs; it is the foremost site for *Didymoceras*, as well as *Hoploscaphites*. Enormous quantities of *Baculites*, *Sciponoceras*, and other heteromorphs are present alongside planispiral ammonites such as *Placenticeras* (see Chapter 8). Their biofacies in the Pierre Shale provide a very clear glimpse into how this environment changed over time.

As discussed in Chapter 8, the WIS was made possible by several things: an extreme greenhouse climate with almost no polar ice, stimulated by extremely rapid seafloor spreading as the supercontinent of Pangea broke up. This led to the high global sea level of the Upper Cretaceous Period. This shallow inland sea offered the vulnerable shells of heteromorphs safe, calm waters. Ammonites in the seaway are often great indicators of time, as well as environmental conditions.

The open ocean realm of the WIS was home to countless marine reptiles such as mosasaurs, plesiosaurs, plus large, powerful fish. While its seafloor was never more than about 50 m deep, it would have been largely inhospitable to most animals due to low oxygen in the bottom waters that encompassed much of the area of the WIS. However, oxygenated oases prevailed around methane seeps on the sea floor, and these were the hubs of much biodiversity for some of the more bizarre-looking heteromorphs. All of the ammonite fossils found in these locations have benthic or epibenthic stable-isotope ratios for at least part of their lives. The locations of these seeps can often be estimated by the faunal presence alone. When only inoceramids and one or two species of pelagic-dwelling ammonites are found, there was not a prolific benthic community. In places where diverse ammonite fauna is found together with other benthic mollusks, there would be a complex ecosystem of benthic invertebrates.

Yezo Supergroup, Island of Hokkaido, Japan

This is the source of many Japanese heteromorphs. Nearly all members of the genus *Nipponites* are found here, along with their close kin. There is some similarity to ammonoids across the Pacific, such as those in the Matanuska Valley of Alaska and at the north end of the Pierre and Bear Paw formations—a shallow sea along the Bering land bridge connected the ammonite faunas of eastern coast of Asia and North America, leading to these closely related faunas.

Heteromorph Families

Common Heteromorphs

Baculitids and scaphitids (discussed in Chapter 8) are usually the most common heteromorphs and were sometimes so prolific that they may have been the most abundant ammonites altogether. As a result, many collectors and non-ammonite specialists in paleontology forget that these two families are, in fact, heteromorphs. Both families speciated rapidly, making them some of the best index fossils of all time. Many of them do not show clear preferences for water quality or position in the water column, with some actually changing their environment through the life cycle. Individual species of baculitids and scaphitids can be found on multiple continents.

While the general bifid shape of their saddles and lobes is present as descendants of the lytoceratid line, both *Baculites* and *Scaphites* drastically reduced complexity of their sutures. This may have been due to their typically small whorl sections, or to sped-up life spans. Either way, it is clear that neither group had need for very much control over the fluids and gasses inside their chambers relative to larger and more complex heteromorphs.

Diplomoceratidae

Diplomoceratids ("double horned" shells in Greek) are members of the larger superfamily Turriloidea (Figure 9.5). Even though their shells may

Figure 9.5 **Nektic and demersal diplomoceratids together under the waves.**

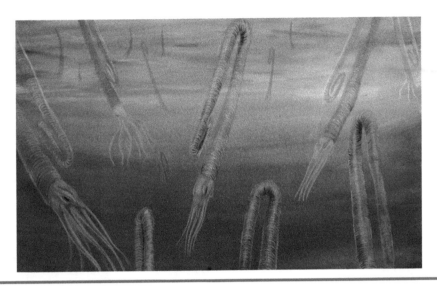

Figure 9.6   A shoal of Antarctic *Diplomoceras maximum,* which floated in open water and hung down from their shells.

appear simpler than some of the nostoceratids, they probably evolved from the spring-shaped nostoceratid *Eubostrychoceras.* Their shells often resembled a paper clip, although they vary in their interpretation of the paperclip form. They had a diverse range of life habits and are found in nearly every heteromorph fossil locality: they vary widely in their sizes, manners of coiling, and especially in their lifespans. We believe that many diplomoceratids floated just below the water surface. Some of them likely filter-fed; others may have had the capacity for preying on larger animals (Figure 9.6).

Scalarites

Some of the earliest diplomoceratids had a circular hook in the beginning life stage, but over time took a number of other shapes later in ontogeny. Some were more elliptical; others formed bizarre triangles. As adults, most of the primitive diplomoceratids hovered above the sea bottom in their environments facing up (Figure 9.7).

Diplomoceras

The most famous of all the diplomoceratids is *Diplomoceras maximum,* the 2 m Antarctic giant on display in the Museum of the Earth at the Paleontological Research Institute in Ithaca, NY (Figures 9.8 and 9.9).

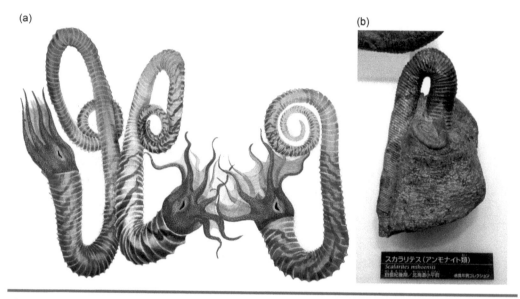

Figure 9.7   (a) *Sormaites teshioensis* and (b) *Scalarites* sp.

Figure 9.8   *Diplomoceras maximum* on display at the Museum of the Earth, Paleontological Research Institution.

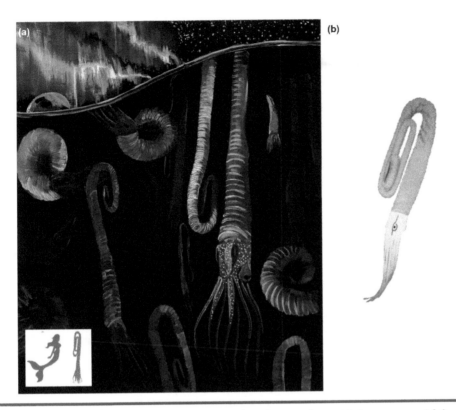

Figure 9.9a ***Diplomoceras maximum* under the Southern Lights, mermaid for scale.**
Figure 9.9b ***Diplomoceras cylindraceum* reconstruction.**

Though the Antarctic continent during the Cretaceous Period was close to its current position over the South Pole, its environment at the time was drastically different from what it is today. Atmospheric carbon on earth was more than twice what it has been in recent millennia, possibly as high as 2,000 ppm (compared to 415 ppm today), causing any given point on earth to be an average of 10°C warmer than it would be today. This warming did not *only* expand the territory of epicontinental seas, but entire temperate forests of fossil trees have been found in Arctic and Antarctic terrestrial deposits of similar age. Under Antarctica's modern cover of ice and snow, the remains of thriving terrestrial and ocean ecosystems lie buried.

The adult *Diplomoceras maximum* floated head-down in the epipelagic zone above the deep waters surrounding Antarctica. The isotopic compositions from nearly all shafts of these elongated shells are so depleted in carbon-13 that it *appears* certain the animals that once inhabited them drifted just below the surface. However, the Southern Ocean was the same temperature from the surface down to 500 m during the Cretaceous, so surficial isotopic ratios would be possible at surprising depths. Along with its size, *Diplomoceras maximum* is puzzling in its lifecycle. *Diplomoceras maximum* is among the first cephalopods to venture into cold water, which seems to have resulted in it having a slower metabolism. In general, ammonoid lifespans have long been estimated between 4 and 35 years—with most of their life expectancies somewhere between the coleoids, which mostly live for 1–2 years, and nautiluses, whose life expectancy in the wild is about 20 years. Using isotope data, the life expectancy of *Diplomoceras maximum* has been estimated at roughly 200 years. Each year, a new growth ridge was added to the shell. These slow-growing cephalopods lived in plain view in open waters, and that some of them managed to survive for centuries are truly astonishing. Today, we see a less extreme example of this with certain abyssal octopodes, which sometimes live for 4 or 5 years, more than twice the lifespans of their closest cousins, while brooding their nests. As ectothermic animals, their ability to live longer—or more accurately, live slower—comes from their frigid surroundings.

Polyptychoceras

As with its cousin *Diplomoceras*, the growth rate for *Polyptychoceras* (Figure 9.10) has been measured using isotopic analysis on the shells' accretionary growth ribs. Some species of *Polyptychoceras* divided their long whorls into as many as four separate planes, or axes, of coiling. Unsurprisingly, then, the life cycle of *Polyptychoceras* appears to be a bit more complex. The living chamber of *Polyptychoceras* is quite long in adult specimens (Figure 9.10), so for much of the animal's life, it would have been a poor floater. Like other "paperclip ammonites," it went through periods of an upward shell orientation, a downward one, horizontal periods, and vertical ones. However, during the downward-facing periods, the animals experienced rapid growth rates; during the upward-facing periods, their growth was slowed. Paleobiologists agree that *Polyptychoceras* spent most of its life near the sea bottom, though it may have floated high in the water column for part of its juvenile period.

The sutures of *Diplomoceras* include the most heavily folded of any ammonite genus. Branching as many as seven or eight times, these incredible complex sutures have been the basis for modeling all sorts of questions regarding highly complex ammonoid septa. Due to its many folds, *Diplomoceras* probably exercised great control over its buoyancy. However, not all diplomoceratids formed paperclips (Figure 9.11). In contrast, its cousin *Polyptychoceras* had sutures that bore similarities to the vertically migrating scaphitids, but it extremely negative buoyancy due to its small phragmocone (Figure 9.12).

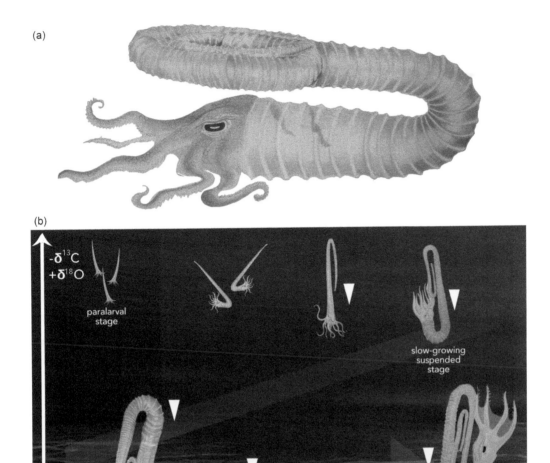

Figure 9.10   (a) Reconstruction of subadult *Polyptychoceras*. (b) Ontogenetic cycle of *Polyptychoceras*.

Ptychoceras

For several decades, *Ptychoceras* was believed to be confirmation of an endocochleate heteromorph, or one which had internalized its shell. In all cephalopods, new shell material is secreted by the mantle, and if an ammonoid were to exhibit endocochleatism, it would suggest the presence of "secondary shell." Secondary shell occurs when the interior

Figure 9.11   **Reconstruction of *Glyptoxoceras*.**

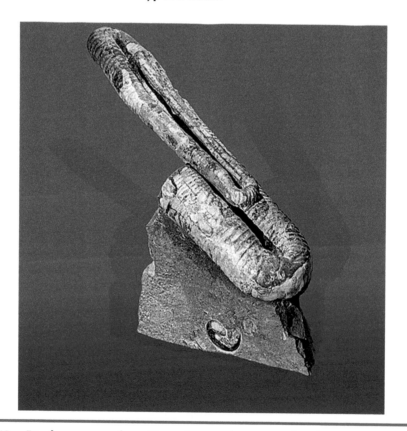

Figure 9.12   ***Ptychoceras* sp. from Yezo, Japan, demonstrating four axes of coiling during ontogeny.**

nacreous layer (the iridescent "mother of pearl" layer) is secreted on both the inside and the outside of the shell instead of just the inside. Internalized shells are known from some of the Ordovician nautiloids, and in some cases, the calcitic counterweighting of Paleozoic nautiloids may have sealed distal parts of the siphuncle, allowing them to break off older parts of the shell no longer in use so that the active chambers could be enveloped by the creature's soft mantle. Today, most coleoids, with the exception of the female argonaut octopodes, have completely internalized their shell. In a cephalopod endoconch, the layer that is traditionally the most exterior is sandwiched between two outward-facing layers of interior shell. However, because *Ptychoceras* only appeared to exhibit this phenomenon partially, it was thought that only part of the shell had been internalized.

The entire possibility of endocochleatism was rejected for *Ptychoceras* in 2020 due to a pathologically wrinkled pattern radiating from the venter of the shell, giving the appearance of an outer secretion point. However, nacre was not present on the shell's outer layer, and it was determined that the periostracum—the usual outermost chitinous layer that gives many shells an almost papery appearance—had enclosed the wrinkles, and therefore, the shell is believed to have been external.

The elliptical whorls, at first glance, appear similar to *Polyptychoceras* and *Diplomoceras* specimens, but a closer look reveals that the axis of coiling changes *four times* (Figure 9.12). Adding convolution, both physically and metaphorically, to the bizarre conch shapes of these creatures, it is apparent that buoyancy control has been revolutionized once again.

Oxybeloceras

Oxybeloceras (Figure 9.13) is an unusually tiny heteromorph. Its phragmocone forms a tiny circle, and then a very long shaft. Superficially, the animal looks very different from other diplomoceratids, but the length of the final bands of growth after the shell makes its signature 180-degree U-turn is where the paperclip becomes apparent.

Another unique trait of the shell of *Oxybeloceras* is the double obtuse angle in the middle of both halves of the shell that would otherwise appear straight. This adaptation may assist in rolling gas between chambers for rocking motions during feeding. Though *Oxybeloceras* is known primarily from the WIS, a single instance of it has been recorded from the New Jersey Atlantic Coastal Plain. The fossil, part of the MAPS collection,

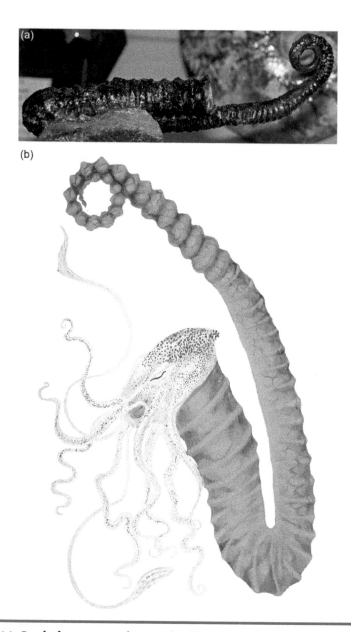

Figure 9.13   (a) *Oxybeloceras meekanum* fossil. (b) Reconstruction of *Oxybeloceras meekanum.*

has not been preserved as a shell. Instead, its cast is preserved in the side of *Exogyra costata*, a large oyster which was abundant in the Cretaceous. These oysters commonly anchored around sedentary objects, including hard detritus, old shells, and even each other, leaving an impression of the anchor object grown around by the oyster. Such casts are known as xenomorphs.

Nostoceratidae

The nostoceratids are famous for the most extremely aberrant shells. Though their simplest form, *Nostoceras* (Figure 9.14), features a shallow coil with a simple, unornamented hook, more bizarre shapes evolved. These ammonites can become very elaborate in their helices, hooks, curls, oxbow bends, and ornamentation (Figure 9.14). Often, multiple very different phases of growth exist over the ontogeny of just one nostoceratid. They can be found worldwide but are especially abundant in Upper Cretaceous beds of North America and Japan. Their unique biodiversity makes them one of the greatest mysteries in ammonoid paleobiology. Unlike planispiral ammonoids and the more simply shelled baculites and scaphites, nostoceratids were probably poor swimmers and had restricted geographic ranges.

Intraspecific variation is taken to an extreme with many of the nostoceratids, alongside many indications of sexual dimorphism. The tightness of a coil, arrangement of nodes or spines, and overall size can vary dramatically within a single nostoceratid species, such that their classifications are often still somewhat disputed. To make things even more complicated, the helical nostoceratids often coil in both directions. Clockwise coiling is described as *dextral*; counterclockwise coiling is described as *sinistral*. It is not known why a given species may contain both sinistrally and dextrally coiling individuals—given that it has been recorded in both micro- and macroconchs, but it appears not to be a secondary sex characteristic. Due to the prevalence of pathologically abnormal septae, suture geometry is generally seen as a last resort for taxonomic assignment, but in many nostoceratids, variation is so prevalent that other, usually more reliable diagnostic methods are insufficient.

Many nostoceratids appear to exhibit evidence of pseudoplanktonism, a life habit in which animals cling to a piece of floating organic debris such

(a) (b) (c)

Figure 9.14   (a) *Nostoceras draconis* reconstruction, (b) *Ainoceras* reconstruction, and (c) *Anaklinoceras reflexum* reconstruction.

as driftwood or kelp. The primary "evidence" for proponents of the pseudo-planktonic interpretation is that their shells are often so complicated that it is unlikely that nostoceratids could swim or control their motion very well. However, it is unlikely that most nostoceratids were actually pseudoplanktonic. Very few of them show evidence for it on their shells, such as indentations of where their shell may have grown around—or at least in contact with—a floating object. Additionally, seaweed is notoriously terrible at becoming fossilized and is seldom found in the fossil record. While marine algae had certainly already evolved on earth, it is unknown how prevalent seaweeds were in the Cretaceous Period, much less alongside ammonites, as they are not found fossilized together.

Didymoceras

The members of *Didymoceras* (Figures 9.15 and 9.16) are likely the descendants of a *Nostoceras* ancestor and are characterized by a nodular helix, usually with some type of dramatic hook at the apical end. All species vary in sinistral and dextral forms. Many of the most common *Didymoceras* come from the WIS, with only a handful of species known in Japan and Europe. The Atlantic Coastal Plain of the eastern United States has also produced *Didymoceras* fossils of the same species found in the WIS, though these appear to be dwarfed relative to their larger western counterparts.

Because these shells tend to go through three to four distinct growth phases over ontogeny (Figure 9.17), it is likely that the animals spent different parts of their lives in age-segregated niches. These niches may correspond to food source, water depth, mode of physical activity, and social groups. The size of the living chamber relative to the phragmocone also changed throughout the animal's lifestyle, each time likely relegating it to a slightly different life habit. Juvenile *Didymoceras*, like most ammonoids, were probably passive residents of the free-water column. Most *Didymoceras* hatchlings would have looked like small *Baculites*, with a shell that was more or less a straight cone, and they seem to have lived planktonically as well. Adults would be found in benthic communities, possibly as scavengers or ambush predators of small prey.

It was commonly thought that *Didymoceras* had to be a passive passive filter-feeder due to its shell, which certainly made swimming a challenge. However, the aptychi of *Didymoceras nebrascense* (Figures 9.18 and 9.19) bears similarity to the beaks of neocoleoids who feed off of large crustaceans and fish. The beak is sharp and dramatically pointed, putting the adult

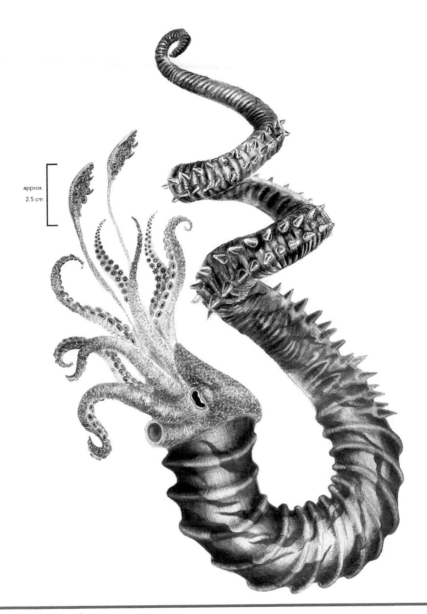

approx.
2.5 cm

Figure 9.15 ***Didymoceras cheyennense* reconstruction.**

Didymoceras in line with other cephalopods who pull apart prey animals that are large relative to their own body size. Furthermore, SEM images showed that they possessed an additional layer of calcite only exhibited by a handful of ammonoids. This has led researchers to believe that *Didymoceras* had the ability to bite into things that were larger than zooplankton.

The middle of the Pierre Shale, where the crustaceans and the majority of *Didymoceras* are found, is dominated by dark shales that indicate

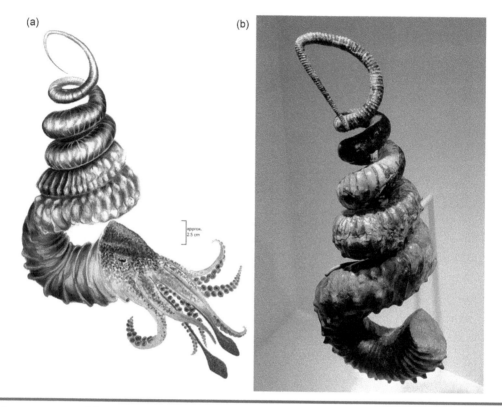

(a)   (b)

approx.
2.5 cm

Figure 9.16   *Didymoceras stevensoni.*

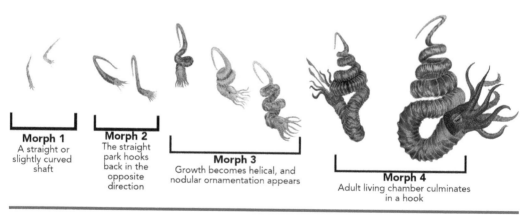

Morph 1
A straight or
slightly curved
shaft

Morph 2
The straight
park hooks
back in the
opposite
direction

Morph 3
Growth becomes helical, and
nodular ornamentation appears

Morph 4
Adult living chamber culminates
in a hook

Figure 9.17   The four ontogenetic morphs of *D. stevensoni*: the straight juvenile, the bent juvenile, the helical subadult, and the hooked adult.

Figure 9.18　**The aptychus *in situ* of *D. nebrascense*.**

the deepest portion of the WIS during the Campanian stage. This deepening was caused in part by the high temperatures during the Cretaceous, and partly by global tectonic events that forced the migration of seawater onto land. Though much of the Pierre Shale represents the deepest facies of the WIS, the sea itself was never truly deep and generally never exceeded 50 m. At this depth, the only light visible would be blue. *Didymoceras* likely managed to catch prey with only the slightest movements of its body, and virtually no movement involving its elaborate shell (Figures 9.20 and 9.21). It spent most of its adult life floating just inches above the sea floor in this dark blue light.

approx.
2.5 cm

Figure 9.19   *Didymoceras nebrascense.*

Figure 9.20   *Didymoceras stevensoni* above seafloor.

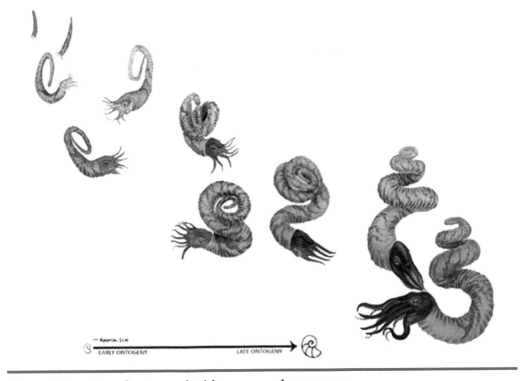

Figure 9.21 **Growth stages of *Didymoceras cheyennense*.**

In all ammonites, it is expected that suture geometry accrues complexity, by adding new levels of subdivision, with each new septa added over ontogeny. However, *Didymoceras* took this to a new extreme. The saddles and lobes of most heteromorphs are Y-shaped (Figure 9.22), taking after their lytoceratid ancestors. In addition to new folds, the lobes and saddles are lengthened over ontogeny. This lengthening is most apparent in the number of new folds added in the bases of the lobes. In Figure 9.23, the stark difference in the shape of the lateral lobe over the ontogeny is shown for *Didymoceras cheyennense* (Figure 9.23).

Pravitoceras

Although it appears to be one of the most conservative heteromorphs, this ammonite is most closely associated with *Didymoceras*. In fact, in the ammonite biozones of Japan, the *Pravitoceras sigmoidale* (Figure 9.24) biozone sits directly above the *Didymoceras awajiense* (Figure 9.25) biozone. Like *Didymoceras*, *Pravitoceras* had a powerful lower beak similar in size, shape, and proportion to that of its *Didymoceras* cousins, and is found in fine-grained sediment associated with calm waters exceeding depths of 40m. For a long time, *Pravitoceras*

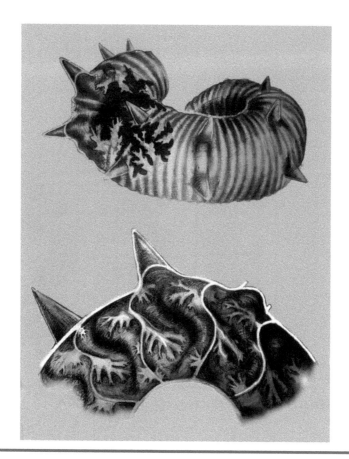

Figure 9.22   **Cross section of *Didymoceras cheyennense*.**

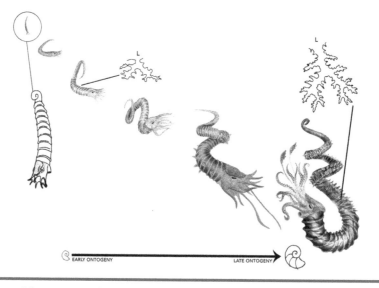

EARLY ONTOGENY   LATE ONTOGENY

Figure 9.23   ***Didymoceras cheyennense* ontogeny with dramatically different lateral lobes in the juvenile and mature phases.**

Figure 9.24 *Pravitoceras sigmoidale.*

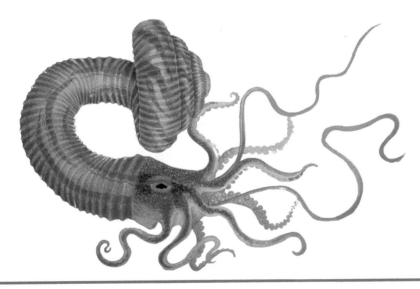

Figure 9.25 *Didymoceras awajiense.*

was believed to be endemic to the Yezo Supergroup, the uppermost Cretaceous unit in Hokkaido which lasts into the early Paleocene, but they are also known from the Izumi Mountains in southwestern Japan. The Izumi beds are slightly older than the Yezo, so both the geographic and temporal ranges of *Pravitoceras* were expanded by their discovery in the Izumi Mountains. *Pravitoceras* is monospecific, meaning that the genus consists of only one species.

Pravitoceras has three morphs. The first is a closed helical, or torticone morph. This morph is small, and difficult to see in lateral photographs. The second and largest morph appears to be a planispiral serpenticone. Like *Didymoceras*, *Pravitoceras* culminates in a large hook at the onset of maturity. The two genera also share two rows of nodes. Due to the its relative flatness and simplicity, *Pravitoceras* would have been much more hydrodynamic in its early life, and drag would increase later in adulthood—the adult pravitocone could move vertically more easily than horizontally (Figure 9.26) and probably evolved for frequent vertical migrations, similar to nautiluses.

It has been widely accepted for decades that *Pravitoceras* is a daughter taxon of *Didymoceras*, but some new research suggests that *Pravitoceras* and *Didymoceras* are one genus. *Didymoceras* first appeared in the WIS, but spread out to northern Japan via the Bering Strait (later called the Bering land bridge during the Pleistocene Ice Age.) Consecutive species in the *Didymoceras* lineage became more tightly coiled—more stout and torticone in their phragmocones—and the final ammonite in this sequence is *Pravitoceras sigmoidale*.

A common feature of *Pravitoceras* is encrusting mollusks and brachiopods. We know that they did not encrust after the ammonoids died and their shells were deposited on the seafloor because the encrusting shells are found on both sides of an ammonite shell. Instead, they hitched a ride on the ammonoids during life. Encrusters, or epizoa, were heavy and caused ammonoids stress. Going back to the Jurassic, bizarre pathological shell growths on ammonites attempted to balance out a heavy encruster. Late *Didymoceras* species, including *Didymoceras morozumii* and potentially, *Didymoceras awajiense*, exhibited encrusters. *Pravitoceras sigmoidale* may simply be a *Didymoceras* that streamlined for vertical migration, to better escape the epizoa.

Bostrychoceras

A thick, helical shell that floated in a vertical orientation, *Bostrychoceras* were most common in Europe (Figure 9.27). The genus *Eubostrychoceras*

Figure 9.26 *Pravitoceras sigmoidale* **ecology, including vertical migrations, hunting of large prey, and epizoa.**

is closely related but is defined as the subset of *Bostrychoceras* that never exhibit nodes.

Eubostrychoceras

In a time when the latitudinal continuity of ammonite distributions was disappearing with the onset of distinctly non-athletic ammonoids, this remarkably successful heteromorph genus managed to achieve global distribution (Figure 9.28). Not only that, but it has been credited with the genesis of the truly bizarre genus *Nipponites*, as well as the entire family of the Diplomoceratidae. In contrast to *Didymoceras*, many species of which

Figure 9.27 *Bostrychoceras polyplocum.*

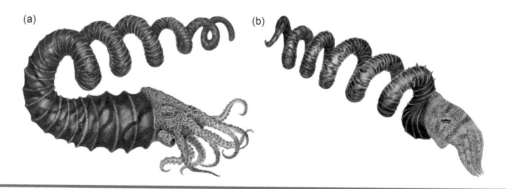

Figure 9.28 *Eubostrychoceras japonicum* sexual dimorphism; (a) Macroconch m, the presumed female; (b) Microconch [m], the presumed male.

adapted for very specific inland sea environments, *Eubostrychoceras* favored relatively shallow waters.

The shells of *Eubostrychoceras* come in many different coil shapes, with some loose helices and some tight torticones. In some cases, determining a *Eubostrychoceras* from other genera like *Nostoceras* or *Didymoceras* has called upon septal geometry. The suture pattern of *Eubostrychoceras* looks almost nothing like a normal nostoceratid. It is lopsided and the characteristic Y-shaped lobes are almost entirely missing.

Muramotoceras

The knotted *Nipponites* is among the best well-known heteromorph, but knotted forms evolved separately several times. *Muramotoceras* exhibit knotting, but only in the juvenile whorls. The adult phase consists of a straight shaft, culminating in a more typical heteromorph hook (Figure 9.29).

Nipponites

By any metric, the strangest of all heteromorphs are members of the genus *Nipponites* (the "stone of Japan"). They were first described from a single specimen of *N. mirabilis* by the legendary Japanese heteromorph expert Hisakatsu Yabe in 1904, and for a long time, they were widely misunderstood. Initially, Yabe and his contemporaries assumed the new, knotted shell

Figure 9.29   *Muramotoceras matsumoto* reconstruction.

was a deformed heteromorph from another species. This was reasonable, given that there was only one specimen known at the time, and its form was incredibly deviant. However, more shells emerged, and writing them off as an isolated pathological incident was no longer possible. From the 1920s, *Nipponites* were believed to have a benthic life habit. The irregularly coiling vermetid snails, marine gastropods which manage to slowly crawl along hard substrate despite their long, curling shells, were brought up as their best living analog. This view held up until the late 1970s, when the first planktonic models for *Nipponites* were proposed by Gerd Westermann and Peter Ward, who noted the shells' small living chambers and low estimated densities. It is now accepted that the overwhelming majority of spaces in these meandering shells was occupied by air—*Nipponites* floated planktonically throughout its lifecycle.

The closest cousins of *Nipponites* are *Eubostrychoceras*, the genus from which most ammonite paleontologists believe *Nipponites* is derived. It is easy to see why: the early ontogeny of both genera is defined by an open gyrocone. *Nipponites mirabilis* is defined by three major ontogenetic morphs: an early gyrocone, followed by its infamous U-bends. The adult shell would then culminate in a hook similar to the ending morphs of *Eubostrychoceras* and *Didymoceras*—though the hook of *Nipponites* was generally more compact. There are three recognized species of *Nipponites*: *Nipponites mirabilis*, *N. bacchus*, and *N. occidentalis*. *Nipponites mirabilis*, the most famous, is found throughout the Turonian and Coniacian marine deposits on the island of Hokkaido. It is the best-known member of the genus, and its shell has been a source of mathematical intrigue for decades. However, in spite of all this variation and the complexity of this manner of coiling, *N. mirabilis* is deceptively ordered. Its growth sequence is mathematically predictable. The middle morph, the U-bends, configures into a spherical mass around the initial gyrocone. When sliced in half, the cross sections of the middle whorls form near-perfect circles, arranged in size and juxtaposition as a logarithmic spiral, mirroring an ordinary ammonite (Figure 9.30).

At first glance, it may seem that there is a great deal of intraspecific variation in *Nipponites mirabilis*, with some tightly coiled, and others more loosely coiled—and heavily ribbed. *N. sakhalinensis*, a subspecies (Figure 9.31) which apparently branched off of *N. mirabilis*, follows the same overall growth trajectory of its parent species but is larger than *N. mirabilis* and bears heavily collared growth bands. True to its name, the variant is only found on Sakhalin Island.

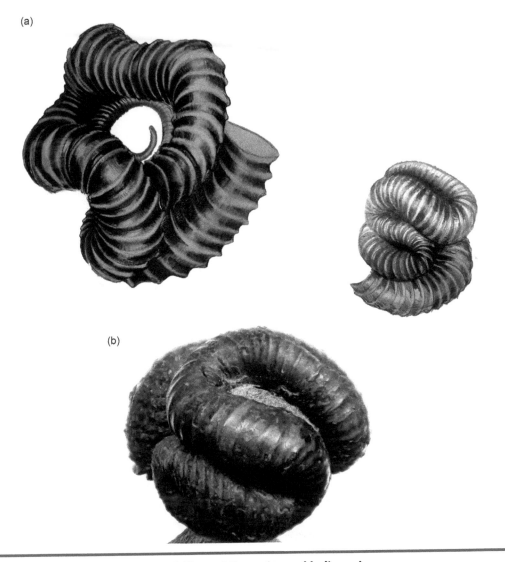

(a)

(b)

Figure 9.30 *Nipponites mirabilis* vs. *Nipponites sakhalinensis*.

Of all ammonites, the greatest intraspecific variation is associated with *Nipponites bacchus* (Figure 9.32). The coils of these ammonites may range from an uncharacteristically loose *Nipponites mirabilis* culminating in a long, straight shaft to bizarre figure-eight forms and just strands of U-bends. Importantly, these ammonites were not sedentary forms that created impressions on other shells called xenomorph, and their unique individual forms were also not the result of pseudoplanktonic behavior. Instead, they floated planktonically, generating shell at angles seemingly chosen at random.

Figure 9.31   ***Nipponites mirabilis*** reconstruction.

Figure 9.32   ***Nipponites bacchus.***

Figure 9.33 *Nipponites occidentalis.*

The only species of *Nipponites* not found in Japan is *N. occidentalis* (Figure 9.33), a species endemic to the Pacific Northwest of the United States, which superficially appears to be the most conservative *Nipponites*. In profile, it appears to simply be an open planispiral made of shallow U-shaped segments. It is actually the last surviving member of the genus.

Hyphantoceras

Meanwhile to the east, a different helical nostoceratid evolved a completely separate vermiticone lineage. *Hyphantoceras* (Figure 9.34) is a bizarre ammonite in that even within the same species, coiling styles can vary dramatically. However, other attributes characteristic to these ammonites were somewhat unchangeable. *Hyphantoceras* were often tiny ammonites, seldom exceeding 10 mm in length, with proportionally quite thick siphuncles relative to the whorl cross section.

In European Mesozoic stratigraphy, *Hyphantoceras reussianum* is an extremely important index fossil. A new ammonite fauna—including large puzosiinae and a roster of heteromorphs—appears very suddenly at the Turonian interval, and *Hyphantoceras* in particular occurred in such great abundance that the sudden appearance of myriad new ammonites is sometimes called the "*Hyphantoceras* [*reussianum*] Event." Ammonite assemblages associated with the "Event" span England to Poland in the north and Spain to Kazakhstan in the south.

Figure 9.34 *Hyphantoceras reussianum.*

The genus *Madagascarites* (Figure 9.35) is sometimes mistakenly viewed as would-be member of *Nipponites*. Only a handful of specimens have been found, and they are found an ocean away from the *Nipponites* species in Madagascar. *Madagascarites* is ornamented with nodes, whereas every other *Nipponites* (as is diagnostic of its sister genus *Eubostrychoceras*) bears only plain growth ridges. Like *Hyphanthoceras*, the shell of *Madagascarites* begins with a straightened cone, unlike *Nipponites*.

Regardless of their respective lifespans or populations, nearly all heteromorph species had temporal ranges between one and two million years. Their use as index fossils is derived from the property of heteromorphs by which they succeed, go extinct, and are quickly replaced by another, closely related group. Their high degree of specialization meant that most heteromorph lineages were not long for the world, but the simplest and most adaptable forms were able to survive the event that did in the dinosaurs. Though vermiticones and trianglicones never appeared in the fossil record again, primitive heteromorphic forms have shown up again and again in diverse cephalopod lineages. The gyrocone resurfaced in the form of the internalized shell (*gladius*) of *Spirula*.

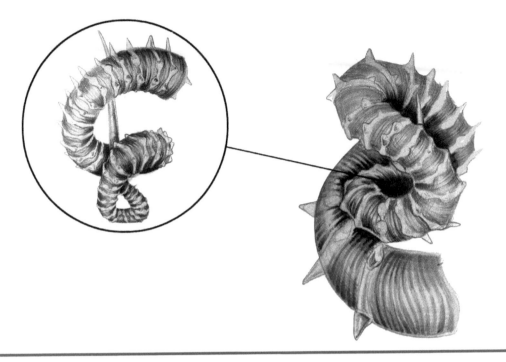

Figure 9.35 **Madagascarites ryu.**

Turrilitidae

Turrilitids, like their type genus *Turrilites* (Figure 9.36), usually feature tightly coiled helices. Though these ammonites resemble snails, their functional morphology is still largely mysterious. It may have been a defense mechanism by which, free of the drag imposed upon open coils, a jet of water from their hyponome would cause a turrilitid to rapidly spin away, startling and confusing its hunter (Figure 9.37). Turrilitids sometimes developed spiky ornamentation, such as is seen in *Ostlingoceras* (Figure 9.38).

Figure 9.36   *Turrilites costatus.*

Figure 9.37   **Drag on helical shells spinning during jet propulsion increases with looseness of the coil.**

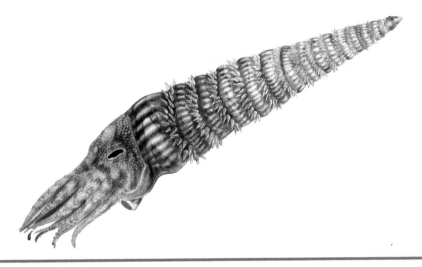

Figure 9.38   *Ostlingoceras reussianum* reconstruction.

Further Reading

Hoffmann, R., Slattery, J.S., Kruta, I., Linzmeier, B.J., Lemanis, R.E., Mironenko, A., Goolaerts, S., De Baets, K., Peterman, D., Klug, C. (2021). Recent Advances in Heteromorph Ammonoid Palaeobiology. *Biological Reviews*, 96(2), 576–610.

Kruta, I., Landman, N., Rouget, I., Cecca, F., Larson, N. (2010). The Jaw Apparatus of the Late Cretaceous Ammonite *Didymoceras. Journal of Paleontology*, 84(3), 556–560.

Peterman, D.J., Ritterbush, K.A., Ciampaglio, C.N. et al. (2021). Buoyancy Control in Ammonoid Cephalopods Refined by Complex Internal Shell Architecture. *Scientific Reports*, 11, 8055.

Peterman, D., Yacobucci, M., Larson, N., Ciampaglio, C., Linn, T. (2020). A Method to the Madness: Ontogenetic Changes in the Hydrostatic Properties of *Didymoceras* (Nostoceratidae: Ammonoidea). *Paleobiology*, 46(2), 237–258.

Seilacher, A. (2013). Patterns of Macroevolution through the Phanerozoic. *Palaeontology*, 56, 127301283. doi: 10.1111/pala12073.

Sessa, J., Larina, E., Knoll, K., Garb, M., Cochran, J., Huber, B.T., Macleod, K.G., Landman, N.H. (2015). Ammonite Habitat Revealed via Isotopic Composition and Comparisons with Co-occurring Benthic and Planktonic Organisms. *PNAS*, 112, 15562–15567.

Slattery, J., Clementz, M.T., Johnson, M.R. (2007). Habitat of the Cretaceous Heteromorph Ammonoid Didymoceras in the Western Interior Seaway. *Geological Society of America*, 39. Abstracts with Programs.

Slattery, J., Harries, P., Sandness, A. (2012). A Review of Late Cretaceous (Campanian and Maastrichtian) Heteromorphic Ammonite Paleobiology, Paleoecology, and Diversity in the Western Interior of North America; in: Cavigelli A. (ed.), *Invertebrates: Spineless Wonders, 18th Annual Tate Conference*. Tate Geological Museum, Casper College, Casper, WY, 76–93.

Tanabe, K. (1983). The Jaw Apparatus of Cretaceous Desmoceratid Ammonites. Palaeontology, 26, 677–686.

Tanabe, K., Kruta, I., Landman, N.H. (2015). Ammonoid Buccal Mass and Jaw Apparatus; in Korn D., De Baets K., Mapes R.H., Klug C., Kruta I. (eds.), *Ammonoid Paleobiology: From Anatomy to Ecology.* Springer, Dordrecht, 429–440.

Tanabe, K., Landman, N.H. (2002). Morphological Diversity of the Jaws of Cretaceous Ammonoidea; in Summesberger H., Histon K., Daurer A. (eds.), *Gabhandlungen Der Geologischen Bundesanstalt*, 57, 157–165.

Tanabe, K., Tsujino, Y., Okuhira, K., Misaki, A. (2015). The Jaw Apparatus of the Late Cretaceous Heteromorph Ammonoid *Pravitoceras. Journal of Paleontology*, 89, 611–616.

Tourtelot, H., Rye, R. (1969). Distribution of Oxygen and Carbon Isotopes in Fossils of Late Cretaceous Age, Western Interior Region of North America. *GSA Bulletin*, 80, 1903–1922.

Tsujita, C.J., Westermann, G. (1998). Ammonoid Habitats and Habits in the Western Interior Seaway: A Case Study from the Upper Cretaceous Bearpaw Formation of Southern Alberta, Canada: Palaeogeography, Palaeoclimatology. *Palaeoecology*, 144, 135–160.

Westermann, G. (1996). Ammonoid Life and Habitat; in Landman N.H., Tanabe K., Davis R.A. (eds.), *Ammonoid Paleobiology, Topics in Geobiology.* Plenum Press, New York, 13, 613–625, Figures 5–6.

Chapter 10

Extinction of the Ammonoids: What Humans Can Learn

Night Comes to the Cretaceous

Until very recently, scientists thought that ammonites vanished at the very end of the Cretaceous (Figure 10.1). But now we know that the last ammonites did not go quietly into the Late Cretaceous night, and actually, they lingered well into the next morning of the Paleocene. For most of the past two centuries, we did not always know that ammonites survived the K/Pg extinction. The fossil record of the ammonites seemed to fade out with the last few million years of the Cretaceous, and no specimens were known from rocks younger than the Cretaceous. To many scientists, the extinction of the ammonites was an indelible marker of the end of the Cretaceous on land, just like the extinction of the non-bird dinosaurs was the signature of the end-Cretaceous extinction on land.

In the 150 years since the Cretaceous was named, and it was clear that Cretaceous dinosaurs and ammonites were no longer alive, people speculated on what killed them off. Most of the ideas were purely speculation with no way to critically test them with scientific evidence. These silly and mostly misguided ideas largely focused on the dinosaurs (too hot, too cold, mammals ate their eggs, and so on) and did not account for the fact that the extinction was a global event in both the oceans and on land, from the plants at the base of the food chain on up—the dinosaurs are just an afterthought. Then in 1978, everything changed. A geologist named Walter Alvarez was studying the rock sequence near the town of Gubbio in the

DOI: 10.1201/9781003288299-10

Figure 10.1 Cretaceous *Pachydiscus* ammonoids as asteroid fragments rain down from the sky.

Apennine Mountains of Italy when he and his colleagues found that the layers of richly fossiliferous Cretaceous limestones full of diverse planktonic microfossils came to an abrupt end, followed by a thin clay layer, and then were overlain by Paleocene limestones which had fewer and less diverse planktonic microfossils (Figure 10.2). Alvarez was puzzled as to what might have caused the sudden shutoff of the limestone production, then deposited the thin clay layer, then was followed by a different kind of limestone.

He took samples of the limestones and clay layer and posed the question to his father, Nobel Prize-winning physicist Luis Alvarez of University of California Berkeley. They decided to look for rare trace elements that might be left from the steady rain of cosmic dust from space, focusing on the extremely rare element known as iridium. If the clay layer had been deposited quickly, there would be very little cosmic dust and almost no iridium. If it had slowly accumulated over tens to hundreds of years, there would be more cosmic dust, and the iridium would be more abundant. They gave their samples to Frank Asaro and Helen Michel, who analyzed them using the nuclear reactor at the Lawrence Berkeley National Laboratory. When the results were in, the iridium concentrations were hundreds of times what they might have expected if the clay were just a product of slow rain of sediments from the ocean surface. For the next year, Luis Alvarez searched for

Figure 10.2   (a) The Cretaceous-Paleogene Boundary in Gubbio, Italy, with Walter Alvarez (right) and Luis Alvarez (left) pointing to the K/Pg boundary clay layer between the limestones. (b) A close-up of the Cretaceous-Tertiary Boundary showing the iridium-rich clay layer in the middle; Italian lira coin for scale.

something that would explain the huge concentrations of iridium, and the others would shoot down his wild ideas. Finally, they hit on the idea that an asteroid had hit the earth, and with it released huge amounts of iridium from the remnants of this extraterrestrial body. They finally published their ideas as Alvarez, Alvarez, Asaro, and Michel in 1980, and it has become one of the most influential papers ever published in the history of geology.

The 1978 discovery of an asteroid impact was met with intense controversy. To those aware of the evidence, the apparent asteroid collision dividing the Cretaceous from the Cenozoic was clearly linked to the total extermination of the non-avian dinosaurs. By extension, the mass extinction borne of this impact was assumed for years to have directly and immediately killed every organism whose fossil record stops even remotely close to the boundary in rock strata separating the Cretaceous and the Paleocene. News of the impact's ruinous effects on life struck the paleontology community and soon after, fanned out across public consciousness. Though extinctions were triggered in communities both on land and in the sea, life in the ocean was surprisingly resilient.

[Originally the boundary between the Mesozoic and Cenozoic was called the "Cretaceous/Tertiary boundary," or K/T boundary, and that was its abbreviation in the early stages of the debate from the 1980s and 1990s. But recently geologists have decided that the old term "Tertiary," for the first 64 million years of the Cenozoic, is outdated and obsolete, and have replaced it with "Paleogene" for the first three epochs of the Cenozoic: Paleocene,

Eocene, and Oligocene. That is why you will find the boundary abbreviated by both "K/T" and "K/Pg" depending on how recent the article is.]

After correlating the mysterious iridium layer at localities around the globe to the Cretaceous-Paleogene Boundary, confidence that a celestial impact was responsible for the ensuing mass extinction built rapidly. However, without the remains of an actual asteroid to pin it on, a mystery remained. Where on Earth did this object meet its end, and how large must it have been to have wreaked worldwide havoc?

In 1990, the solution was found buried deep underneath the jungles of northern Yucatan in Mexico. At a spot named Chicxulub ("CHIK-zoo-loob"), geophysical evidence and later drilling showed that there was a gigantic crater beneath the surface, filled in with debris and then hundreds of feet of Cenozoic sediments. The Chicxulub asteroid is estimated to have been about 10 km (about 6.2 miles) wide. Though the width of this asteroid is only about the length of Manhattan Island, the effects of a celestial object colliding with Earth are generally much more widespread than the object itself. This is for several reasons. Objects entering Earth's atmosphere burn up. As a result, meteorites that reach the ground tend to contain only their absolutely most durable components, which could not be destroyed during this ignition— mostly iron, nickel, and occasional olivine. Objects also gain velocity as they become enveloped by Earth's gravitational field, and once they have been pulled into a reasonable proximity to the planet, they accelerate with a vengeance. Of course, larger objects are heavier, and the effects of gravity on them are greater. Because of the combustion and the force at which objects collide with Earth due to gravity, even a significantly smaller celestial object, 6 or 7 m in diameter, has the potential to destroy an entire modern city. However, the end-Cretaceous impactor, itself the size of a city, had significantly greater ramifications. At the height of its acceleration, the asteroid is estimated to have hurtled toward Earth's surface at about 30 km/s. The force and heat associated with the collision ensured that, at the moment of impact, land animals as far as 1,700 km away were subjected within seconds to third-degree burns all over their bodies, brutally cooked alive inside a fireball that extended further and faster than any animal could possibly escape it. If humans were subjected to these conditions, their clothes would spontaneously burn away. Magnitude-10 earthquakes reverberated across most of the globe within the first hour after impact. (For reference, the most powerful earthquake ever recorded was a magnitude-9.5, in Valdivia, Chile in 1960.)

Though conditions quickly killed most of the organisms closest to the impact, the effects of the event as a global extinction were brought on by the cloud of debris which was instantly kicked up by the force of the collision. This debris, which contained Earth sediment as well as fragments from the asteroid, called *ejecta*, was thrust just far enough out of the earth's atmosphere to re-enter. As we know, objects which enter Earth's atmosphere catch fire. The dense ejecta cloud was no exception, and it lit up the upper atmosphere. This thick cloud of burning particulates fell to Earth, igniting widespread wildfires and engulfing any animals that could get low enough to escape its wrath in heavy smoke and burning air. Eventually settling on the ground over a period of weeks or months, the ejecta sediment formed a thin layer of sediment found at the same period of geologic time around the world. This is the iridium layer.

But the biggest killer of all was not the impact itself, nor the debris that rained down. It was the huge volumes of dust-sized particles blasted into the stratosphere by the impact. This remained up in the stratosphere for weeks to months, creating a dark cloud of dust that blocked most of the sunlight and plunged the earth into a "global winter." The longer the darkness remained, the more it wiped out most plant life in both the oceans (phytoplankton) and on land with the Cretaceous trees and bushes. This destruction of the plant life at the base of both the marine and terrestrial food pyramids was the primary killer. As the plants withered and died, the zooplankton in the ocean and the herbivorous animals on land slowly starved to death. Soon their predators in the oceans and on land followed suit.

Volcanoes and Sea Level Changes

But is the impact the whole story? If you listen to the popular media, it's always "asteroid killed the dinosaurs, end of story." The media love to give simplistic answers that can be conveyed in a title only, or a short sound bite, and don't like to mention all the complexities that accompany any major scientific controversy. While the immediate effects of the collision were absolutely disastrous, the end-Cretaceous extinction is much more nuanced than the simplistic impact notion held by many people whose greatest paleontological familiarity is with dinosaurs. This is especially true in the oceans. Though the asteroid undoubtedly played a crucial role triggering the series of extinctions which rang through the earth's biosphere, environmental conditions which prevailed through the Cretaceous had already sentenced many organisms to a gradual decline.

Even when the controversy started in the early 1980s, many geologists pointed out that there was another gigantic catastrophe in the latest Cretaceous that was already well known: the eruption of the Deccan lavas in what is now western India and Pakistan. The Deccan lavas were the second largest eruption in all of earth history, exceeded only by the Siberian eruptions that ended the Permian (see Chapter 6). The lava flows are over 2,000 m (6,600 ft) thick, with a volume of about 1,000,000 cubic km (200,000 cubic miles), and originally they may have covered about 1,500,000 square km (600,000 square miles). These could have brought up huge amounts of sulfur dioxide, plus dust and ash in the atmosphere and have a similar effect as the "asteroid nuclear winter" model. They caused a global temperature drop of 2°C (3.6°F) just 25,000 years before the impact happened. And they are dated to the very end of the Cretaceous, only a few thousand to a few hundred years before the impact. Even more interesting: it turns out that mantle-derived volcanoes like Kilauea also have lots of iridium in them. (Iridium is extremely rare in the crust but slightly more abundant in the mantle and in space.)

And if that's not complicated enough, the end of the Cretaceous was also marked by an enormous drop in global sea level. The huge expanse of shallow seafloor of the Western Interior Seaway dried up and vanished, and the same happened to shallow epicontinental seas around the world. The immense area of oceans they represented also vanished, greatly reducing the habitat for most of the marine life, including the ammonites.

So which mechanism is the most important? Now some 43 years since the original publication of the impact hypothesis, the answer is: "It's complicated." Despite the simplistic view you hear in the media, and from some scientists, the controversy has never died down. The meetings of various geological societies, like the Geological Society of America, still have sessions arguing back and forth about whether the impact or the volcanoes or the sea-level drop was the most important. This means there will probably never be a simple "right" answer. We know that the Deccan eruptions were going on well before the end of the Cretaceous, causing severe stresses in the global climate, and we see evidence in both the oceans and on the land of extinctions happening well before the end as well. The weird cone-shaped bivalves known as rudistids (Figure 8.3) and the huge flat inoceramid clams both were declining during the latest Cretaceous and vanished long before the impact, so they were already gone when the rock from space arrived. A distinguished panel of British paleontologists (MacLeod and colleagues, 1997) showed that most marine invertebrates (gastropods, bivalves, echinoids, crinoids, bryozoans, and corals) show only minor extinctions during the time, and their extinctions are protracted through the latest Cretaceous, not all ending abruptly at the iridium layer, so the global climate and global oceans were going through long-term stresses well before the impact.

The record on land also gives mixed signals. Sure, the non-bird dinosaurs died out, but most studies show that they died out gradually through the latest Cretaceous. There were big changes in the land plants, so the spores and pollen of the characteristic latest Cretaceous *Aquilapollenites* flora vanish, and are replaced by fern spores (which are tolerant of cold dark conditions) at the K/Pg boundary. But most other land animals went right through the impact layer with minimal change. There was almost no extinction in the crocodilians and the crocodile-like champsosaurs, even though many were as large as the smaller dinosaurs and unable to adapt quickly to the cold dark conditions of the post-impact world. Nor is there much extinction in the abundant turtle fauna, or any extinction among the frogs and salamanders. This last fact is important because amphibians have porous skins and cannot tolerate much acidity in the water, so if the world were really bathed in sulfuric acid, as some extreme versions of the impact model claim, there would not be a frog or salamander alive today. The mammals show only a minor effect, with the opossum-like marsupials being largely replaced by placental mammals, but there is no mass extinction at the boundary.

In short, both the marine and terrestrial fossil record give a complex answer and show that the end of the Cretaceous was not the simplistic scenario that the media constantly promote. Instead, the evidence shows that the latest Cretaceous was already a hellish world, with the extremes of climate due to the Deccan volcanism, and the loss of marine habitat due to sea level retreat. The asteroid was more like the final blow, the coup de grâce, on what was already a bad time on Planet Earth.

Life and Death in the Oceans

In 1997, Norman MacLeod and a distinguished panel of British paleontologists examined whether the effects of the Cretaceous-Paleogene extinction were felt as immediately by benthic marine communities as terrestrial ecosystems. It turns out that, though a faunal turnover in deep-sea environments did occur, it was on a delay. With a great enough distance between the ecosystem and the water surface, survival could be expected for much of the benthic fauna—if only for a little while. Worldwide, the benthic foraminiferal turnover marks an especially sharp extinction and speciation boundary. Many geologists who work on the K/Pg extinction regard the end-Cretaceous extinction as not one, but several different K/Pg boundaries: an impact-based K/Pg boundary and an organismal K/Pg boundary. The asteroid impact is obviously the first, and it is reflected with shocked quartz, silicate spherules, and a brief period of chaotic sedimentation directly above the iridium anomaly. In many marine limestones, the boundary between Cretaceous and Paleocene foraminifera and mollusks does not occur for several centimeters above the base of the impact layer.

Over the course of several significant papers in the late 1980s, paleontologists William Zinsmeister and Carlos Macellari described a rich assemblage of heteromorph and planispiral ammonites from Seymour Island, which is at the tip of Antarctica's Palmer Peninsula. These specimens appeared right up to and many meters above the iridium layer that marked the K/Pg boundary. The discovery was later interpreted as reworked specimens—ammonites from an earlier time which have been eroded and then redeposited into younger sedimentary rocks. The impact theory of clear-cut, immediate death for all end-Cretaceous ammonites was temporarily shielded from this revelation, and shockingly, both the field and the public forgot the potential implications of these specimens. However, public perception was overlooking a key detail: when the age of an ammonite is sometimes unknown, microfossils are commonly used to determine what time period it belongs to. Foraminifera and dinoflagellate microfossils from inside the ammonite shells found on Seymour Island were extracted from a handful of Seymour Island localities. Ammonite chambers from assemblages at several localities contained dinoflagellates from both the Cretaceous sediment and the Paleocene. This suggested that both the ammonites and the dinoflagellates were largely unaffected by the impact, cohabiting the Seymour Island paleoenvironment for potentially hundreds of thousands of years in the early Paleocene.

Macleod's examinations of benthic paleoenvironments following the impact included foraminifera and bivalves, but it turns out that ammonoids that were adapted for sufficiently deep water could also benefit from protection. There is evidence from several fossil localities that scaphitids may have survived the extinction of other ammonites, if even for a brief time. In 2002, Polish paleontologist Marcin Machalski discovered *Hoploscaphites* ammonites above the iridium anomaly in Denmark's *Cerithium* in Denmark.

In 2004, American paleontologist Ralph Johnson, who has collected and now curates the world's most extensive collection of Campanian and Maastrichtian Atlantic Coastal Plain fossils (the Monmouth Amateur Paleontological Society—or MAPS—Collection), discovered an assemblage of *Discoscaphites iris* sitting in a section of the Tinton Formation, called the *Pinna* layer. The Tinton Formation is the uppermost rock layer of the Cretaceous in New Jersey and contains the iridium layer associated with the end-Cretaceous extinction event, as well as the first several years of the Paleocene. The iridium layer is found at the base of the *Pinna* layer, which gets its name from a characteristic abundance of the pen shell clam *Pinna laqueata*. Everything below this thin layer of ash is Cretaceous; everything above it is Paleocene. Johnson realized these ammonites had been deposited, not during the Cretaceous Period, but in the early Cenozoic. The *Discoscaphites iris* turned out to be the last appearance data for all ammonites in North America. Since the discovery, other ammonites have since been found in early Paleocene sediment of New Jersey. Though a handful of other ammonites occur as well, overwhelmingly, they are scaphites and baculitids.

At first, like the Antarctic heteromorphs, Johnson's discovery was believed to represent reworked specimens. This was the more plausible explanation—that *Discoscaphites* did not live past the Cretaceous; their shells had simply floated for several months or years after the ejecta from the K/Pg impact cooked nearly all life on earth. With only 40% of sea life remaining, there were far less opportunities for their postmortem-drifting shells to be disturbed, leaving more of them for delayed deposition sometime in the early Paleocene. However, several clues exist at this interval which suggest that at least some of the *Discoscaphites* were living for years after Chicxulub. Importantly, *Pinna* clams stand vertically in the substrate (Figure 10.3). The *Pinna* shells interspersed with ammonite fossils were amazingly preserved in life position, indicating that their burial was rapid, but gentle. The *D. iris* found with them are often whole, lessening the likelihood that they were eroded and then reworked. *D. iris* found in later sediments exhibited significant differences in preservation: namely, in concretion nodules surrounded by worm burrows. The nodules, coupled with heavy bioturbation, could mean that the ammonites found higher in sequence were reworked, whereas the ones in the *Pinna* layer were true Cenozoic ammonites.

It is possible that *Discoscaphites*, and every ammonite species that hung on a while after the impact, had no real preference for water depth. In 2021, Rene Hoffmann and others determined that *Hoploscaphites*, a fellow scaphitid and fellow K/Pg survivor, migrated between the pelagic and the

Figure 10.3 **Discoscaphites iris swimming in the Pinna layer.**

benthos periodically for multiple breeding seasons. *Hoploscaphites*, then, can be considered both a demersal and a hemipelagic genus (Figure 10.4). After death, many ammonite shells, like nautilus shells, could likely float indefinitely, their chambered shells filling with gas and keeping them afloat. However, it was determined in the 2010s by Rene Hoffmann and colleagues that smaller shells, such as those of scaphitid ammonites, did not float for very long, or very far, from the point at which the animal died. Other examples of scaphitids in the early Cenozoic emerged, so the Paleocene scaphitids from New Jersey can no longer be discounted as an isolated incident. Specimens from the Brazos River region of Texas may have lived for more than a quarter-million years after the end-Cretaceous impact. *Scaphites* specimens are rumored from Turkmenistan that are dated to 61 million years old. These ammonites, sometimes as much as several million years into the Cenozoic, do not always show evidence of being reworked by erosion and sedimentation.

Figure 10.4 **Mature and subadult *Hoploscaphites* swimming near the surface, in contrast to the benthic placement of the genus at other points in its ontogeny.**

The debate over whether any of these ammonites truly lived into the Cenozoic persisted until 2005, when Machalski described a number of scaphitid and baculitid ammonites from the *Cerithium* Limestone in Denmark. Of the Danish ammonites, a smoking gun for post-Cretaceous survival emerged: a *Hoploscaphites constrictus* (Figure 10.5) whose inner chambers were filled with Paleocene-aged foraminifera. It is not possible to fill the chambers with sediment and then, several million years later, for the

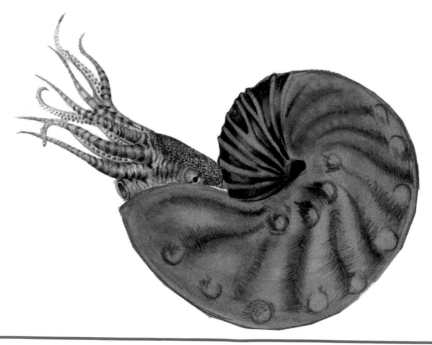

Figure 10.5 *Hoploscaphites constrictus.*

chambers to be emptied of lithified material and then refilled—this would destroy the shell.

Though ammonites as a whole did not survive the K/Pg boundary by more than several hundred thousand years, the few that did share some traits in common. They were primarily ammonites that reproduced rapidly and usually were somewhat adaptable in terms of their position in the water column. Notably, the scaphites appear to be a common denominator. Especially given that some scaphite eggs were laid far from the surface, the 40–50 m of water between them and the chaos above could have shielded them from the ejecta. Baby scaphitids could have lived out their early stages without migrating upward until the perilous aftermath of the impact had passed.

In addition to scaphitids, baculitids seemed to have done reasonably well in the low latitudes. *Baculites* were found in Cenozoic sediments in North America and Europe. Cenozoic *Pachydiscus* specimens were recovered both in New Jersey and Seymour Island. Though Cenozoic ammonite fossils are rare, the widespread occurrences of the same species thousands of miles apart suggest much greater geographic distribution of these species than is reflected in the fossil record. This becomes apparent when we think of how *Hoploscaphites*, a genus that was primarily endemic to the Western Interior

Seaway in the middle of North America, made it to Europe. Hundreds of *Hoploscaphites* shells have been recovered in the central United States and Canada. However, comparatively few have been recovered in Europe. For a while, a theory posited that by some freak accident, a handful of small, relatively defenseless American ammonites successfully migrated across the open Atlantic Ocean (from which recovering fossils isn't possible) to Europe, and somehow won the preservation lottery. Needless to say, the likelihood of this scenario is infinitesimal. For every one fossil that exists, countless unpreserved brethren once roamed alongside it. We must therefore conclude that the entire space between members of a given species was at least part of the range of that species, whether it is possible to recover fossils there or not. That is to say, hemipelagic and nektobenthic ammonites inhabited the world's oceans in large numbers and shocking biodiversity for almost half a million years after the Chicxulub impact.

The notion that several lineages of ammonoids survived for hundreds of thousands of years after a catastrophe of this magnitude begs the question of, if they survived even worse odds in the Devonian and Permian mass extinctions, why they failed to squeak through one last crisis. The end-Permian extinction spared only one ammonite genus, yet biodiversity rebounded to its pre-cataclysm numbers within an unbelievably short amount of time. What environmental conditions prevented not even one of the several surviving ammonite families from persevering through the Cenozoic?

The End of the Ammonites

The Paleocene-Eocene Thermal Maximum (PETM) has been regarded by Cenozoic paleontologists as one of the most significant environmental events since the K/Pg. An extinction event in its own right, sudden intense global warming, and influxes of greenhouse gasses meant that numerous animals which thrived in the absence of dinosaurs did not make it out of the Paleocene. But this event occurred almost ten million years after the K/Pg impact, and several million years had passed between the total extinction of ammonoids and the onset of the PETM.

Paleontologists were eventually inspired by the significance of the PETM to look at how environmental conditions which led to other extinctions may have occurred throughout the Paleocene. It turned out that the Paleocene was dominated by punctuated periods of global warming, spikes in atmospheric carbon, and severe ocean acidification that were on par with the PETM. The first of these disastrous hyperthermal events occurred about 200,000 years after the K/Pg impact in the Danian, which is the earliest stage of the Paleocene. Called the Dan-C2 event for the Danian stage and the high levels of carbon which characterizes it, its sedimentological record can be correlated within tens of thousands of years—the blink of a geologic eye—of the last Cenozoic ammonite.

It is not entirely clear why the Paleocene was so heavily affected by hyperthermal events, but many of them bear striking similarities to one another, and by extension, to the climate crisis we face today. They are overwhelmingly started by a sudden influx of atmospheric carbon. Carbon dioxide can enter our atmosphere from biological processes, from volcanoes, from outer space, and a litany of other sources. When it builds up, it is a formidable greenhouse gas with a less well-known, but nonetheless deadly comorbidity: ocean acidification.

Ocean Acidification

Today, ocean acidification affects coral reefs a variety of marine snails (Figure 10.6), and any organism that makes a shell from calcium carbonate (calcite or aragonite). It also affects the habitable waters of animals without a shell and confuses the seasonal cycles of numerous species. As we discussed, carbon dioxide is a normal part of the environment. The oceans are equipped to absorb a certain amount of carbon dioxide from the atmosphere, and marine life is adapted for carbon entering the ocean—up to a point. However, when that limit is reached, imbalances emerge. When molecules of carbon dioxide enter the ocean, the carbon and oxygen atoms separate. Most of the time, CO_2 bonds with H_2O molecules to form carbonic acid. However, a subset of the carbon atoms find themselves without a partner. Both carbon and oxygen must bond to another atom. For the oxygen constituents of CO_2, this is easy: they can simply bond with one another, forming O_2 molecules. Excess lone carbon atoms are then attracted to other carbon atoms, most of which are already tied up in the calcium carbonate shells of mollusks, corals, and other invertebrates. Because the lone carbon atoms have to form carbonates molecules, they carbonate comes from the only form locally available. Robbed of carbon atoms, the calcitic and aragonitic shells of marine invertebrates are gradually corroded away (Figure 10.6). As acidification and temperature stressors accelerate, self-healing abilities are curtailed in many marine species, compounding the destruction of animals' exoskeletons.

Figure 10.6   Ocean acidification acting on a pteropod, a small planktonic snails.

The time period during which ammonites went extinct once and for all is probably correlated to the onset of the Dan-C2 hyperthermal event, and especially the ocean acidification that came with it. Why, then, did nautiloids survive? Several possibilities have been raised, including the fact that living nautiluses are able to access water depths that would have been inaccessible to ammonites due to their relatively thicker shells. Another possibility is simply that the respective shell thicknesses of ammonites and nautiloids themselves protected nautiluses from the destruction of their shells, whereas the thin shells of ammonites could not hold up against the excess carbonic acid.

An Unrepeatable Event

Evolution is a one-way ticket for lineages of organisms. When a vacancy opens up in the niches of an ecosystem, it is only a matter of time before a group moves in and occupies it. This leads to convergent evolution. Although the same evolutionary pressures may cause the new species to evolve *some* of the same traits as the previous occupant of the niche, convergent evolution never makes a perfect copy of the organism that was lost. There will never be another group of ammonoids again.

That said, it can be argued that evolution tried more than once to remake the ammonoids. The first time in the Jurassic, and then later in the Eocene all the way into the Miocene, completely separate lineages of nautiloids evolved zig-zag sutures that looked strikingly similar to those of a goniatite. Even more confusingly, the sides of the suture appeared to have become convex rather than concave, which is what we expect for ammonoids. By the Eocene, the lateral lobes of these nautiloids had become so dramatically convex that were very difficult to distinguish from the goniatitids which had died almost 200 million years prior.

These nautiloids were not ammonoids; they did not descend from ammonoids and still mostly maintained the siphuncle that passed through the center of the septa. They had no close relation to the goniatitids. Their transitional phases were not particularly comparable to bactritid cephalopods, which did not even make it out of the Triassic. These adaptations were merely part of the most recent pulse in nautiloid biodiversity, which occurred in large part due to the vacant niches left by ammonoids after they eventually died out within half a million years of the end-Cretaceous extinction event (note how much more slowly that pulse came for nautiloids—10–20 million years after the extinction of ammonoids, whereas ammonoids could adaptively radiate in a million years or less). However, zig-zagging nautiloids failed to achieve the level of evolutionary success that would have allowed them to effectively replace goniatitids in relatively recent time.

The closest thing we have today to a living ammonite is also one of the ammonites' closest living relatives. Females of the species *Argonauta argo*—the argonaut octopus, or, more confusingly, the "paper nautilus"—bear striking resemblance to their ammonoid cousins—if only superficially. The female argonaut secretes an external shell. After mating with a smaller male argonaut, the female broods her eggs in this thinly walled calcitic spiral, inside of which she also rests, passively floating through the open ocean until her eggs are ready to fend for themselves. When the time is right, the octopus releases her shell into the ocean with her eggs in it, and hatchlings emerge weeks later (Figures 10.7–10.8).

Though the argonaut is, by any stretch of the imagination, the most similar creature to a "living ammonite" that we can see today, much like every other example of convergent evolution, there are distinct differences between argonaut octopodes and ammonoids. The most obvious difference, of course, is that the octopus is not permanently attached to its shell, and the argonaut shell is without septa or a siphuncle. Instead of using the shell to control buoyancy or movement, the argonaut simply uses the shell as a floating nest that is abandoned after a nominal period of time. Male argonauts do not secrete an external shell. The most intriguing difference between argonaut and ammonoid shells, however, is the mechanism by which they form.

Like all mollusks, ammonoids deposited new portions of the shell at the back of the mantle. Though the mechanism by which they did this is

Figure 10.7 Female *Argonauta argo* octopus with shell (redrawn from Comingio Merculiano 1896).

Figure 10.8 The female argonaut secretes shell with her arm webbing. Photo of *A. argo* with a damaged egg case exposing her eggs. (Courtesy of Bernd Hofmann.)

still the subject of intense debate, it is clear that the rear mantle structure was in some way directly instrumental in the secretion of new septal walls and new chambers. Nautiluses also secrete new shell via the rear mantle structure.

However, argonauts don't do this. For a long time, the point of secretion on argonauts was a mystery, until 2021 when a team of researchers led by Antonio Checa discovered that shell was secreted from the webbing between the dorsal arms of the argonaut. The specific mechanism by which the arm webbing secretes shell is not yet fully understood, but the discovery places argonauts in a category by themselves as the only mollusks to secrete shell from outside the mantle cavity.

Coleoid Takeover

As nautiluses continued, plodding through their vertical migrations, themselves a remnant of countless bygone eras of the planet's history, coleoids (octopodes, cuttlefish and squid) are the dominant cephalopods today (Figure 10.9). The shell-less survivors fossilize far less frequently than their late ammonoid cousins, but without the fateful event which did in the ammonites, we know that niche spaces would not have opened and given rise to the extraordinary biodiversity of living octopodes and squid. Coleoids took the environments and ecological roles available to ammonoids several steps farther: unburdened by an external shell, octopodes and squid were able to expand into countless environments that ammonoids could not occupy. Octopodes and squid developed a mastery of color to which no animal in the history of the world has ever come close. This feat is far more impressive when one remembers that the cephalopods do not see color the way that most humans do. Altering their physical texture, color patterns, and even learning to glow, octopodes and squid colonized the reefs, the open ocean, and even the abyssal plains. Thanks to coleoids, it can no longer be said that intelligence is exclusive to vertebrates. Seemingly endless takes on the basic eight-armed bauplan have come to be.

The Next Mass Extinction?

Despite surviving five mass extinctions, including ones that decimated ammonoids, nautilus numbers are dwindling. Their iconic shells have

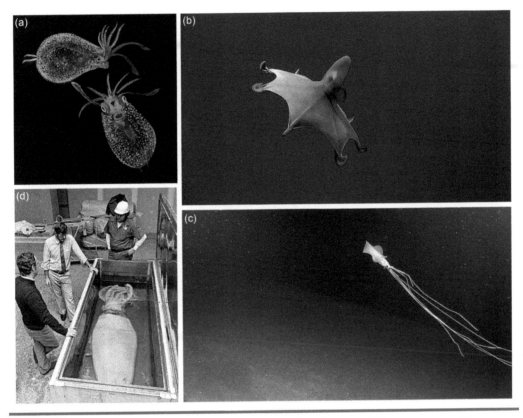

Figure 10.9   Coleoids which specialized into niches inaccessible to ammonoids. (a) Glass squid *Cranchia scabra* by Paul Louis Oudart (1847). (b) Dumbo octopus *Cirrothauma murrayi*. (Courtesy Wikimedia Commons.) (c) Bigfin squid *Magnapinna* sp. (Courtesy NOAA Photo Library.) (d) Giant squid *Architeuthis dux* being studied by Mike Sweeney, Clyde F. E. Roper, and Charles Beggs at the National Museum of Natural History. (Courtesy Wikimedia Commons.)

become a liability, and they are hunted for trinkets at a rate much faster than their small egg clutches can replace their populations.

Like ammonoids, coleoids breed rapidly, lay lots of eggs, and evolve quickly. Unencumbered by a shell, many are not at an immediate risk to ocean acidification from human-induced carbon. Still, it is inaccurate to say that coleoids are unaffected by climate change. Across Europe, octopodes have migrated north in large numbers to escape warming oceans. Cephalopod blood, called hemocyanin, functions best at a specific temperature, and they do not feel well when the water becomes too warm for them.

Sadly, it is coleoids themselves who are at risk to, through no fault of their own, contribute to climate change. Octopus farming is a new trend which has put both octopodes and the environment in a terrible situation. Factory farming octopus is highly controversial. Due to their incredible

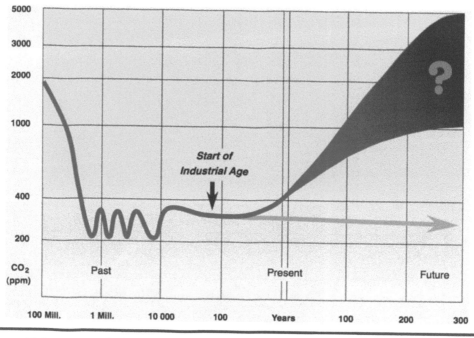

Figure 10.10 **A graph showing atmospheric climate increase versus input of anthropogenic greenhouse gasses. (Courtesy Hannes Grobe.)**

intelligence, the industrial environment is especially cruel. The environment itself will also pay a price for this new practice. All cephalopods, and most animals living in the ocean, are carnivores. They consume fish, and expelling pescatarian waste is well sequestered in the wild, but in an industrial complex, feeding and accommodating the waste of thousands of octopodes create an enormous source of greenhouse gasses, filthy water that must be treated, and a heavy energy expenditure.

Human-induced climate change is accelerating much more quickly than during many of the worst mass extinctions we have discussed, having hurtled toward a similar event in just a few centuries (Figure 10.10). If climate-driven ocean acidification does not subside, our proxy for a last "living ammonite," the argonaut, is set to meet a similar fate to Cenozoic ammonoids. The shell of the argonaut is called a "paper nautilus" for a reason: it is the thinnest mollusk shell of any extant species. The periostracum is a layer of shell found in most mollusks, including ammonoids, and it protects the shell against corrosion and other injuries. Using the shell of the argonaut species *Argonauta nodosa*, a team of researchers led by Kennedy Wolfe found that argonaut shells may not see their species through to the next century if global warming and ocean acidification continue to rise as projected. The paper-thin shell of the argonaut does not possess a periostracum. Within the next five to seven decades, global ocean temperatures and seawater acidity is expected to rise to such a level that significant calcium carbonate mass will be lost from argonaut shells.

The ammonoids were survivors. They survived the great Devonian extinction, and the mother of all mass extinctions in the Permian. They survived Triassic and Jurassic extictions, and a string of smaller incidents throughout the Cretaceous. They probably even could have survived the K/Pg extinction that wiped out the dinosaurs if not for acidification in the early Paleogene. Nature's experiments at making "new ammonoids" occurred much later in the late Oligocene and Miocene, when acidification was no longer a threat. The argonaut, the lone survivor of the post-ammonoid experiments, may not survive long with what the changes we humans are inflicting on the world's seawater, and like them, countless other cephalopods face a changing world, redistribution of their habitats, and an uncertain but dimming future.

Further Reading

Alvarez, W. (1997). *T. Rex and the Crater of Doom*. Princeton University Press, Princeton, NJ.

Alvarez, L.W, Alvarez, W., Asaro, F., Michel, H.V. (1980). Extraterrestrial Cause for the Cretaceous-Tertiary Extinction. *Science*, 208(4448), 1095–1108.

Archibald, J.D. (1996). *Dinosaur Extinction and the End of an Era*. Columbia University Press, New York.

Checa, A.G., Linares, F., Grenier, C., Griesshaber, E., Rodríguez-Navarro, A.B., Schmahl, W.W. (2021). The Argonaut Constructs Its Shell via Physical Self-organization and Coordinated Cell Sensorial Activity. *iScience*, 24(11).

Dingus, L., Rowe, T. (1997). *The Mistaken Extinction: Dinosaur Evolution and the Origin of Birds*. W.H. Freeman, New York.

Keller, G., McLeod, N. (1996). *Cretaceous-Tertiary Mass Extinctions: Biotic and Climatic Change*. W.W. Norton, New York.

Landman, N.H., Cobban, W.A., Larson, N.L. (2012). Mode of Life and Habitat of Scaphitid Ammonites. *Geobios*, 45(1), 87–98.

Landman, N.H., Garb, M.P., Rovelli, R., Ebel, D.S., Edwards, L.E. (2011). Short-Term Survival of Ammonites in New Jersey after the End-Cretaceous Bolide Impact. *Acta Palaeontologica Polonica*, 57(4), 703–715.

Landman, N. H., Goolaerts, S., Jagt, J.W.M., Jagt-Yazykova, E.A., Machalski, M., Yacobucci, M.M. (2014). Ammonite Extinction and Nautilid Survival at the End of the Cretaceous. *Geology*, 42(8), 707–710. doi: 10.1130/g35776.1.

Landman, N.H., Johnson, R.O., Edwards, L.E. (2004a). Cephalopods from the Cretaceous/Tertiary Boundary Interval on the Atlantic Coastal Plain, with a Description of the Highest Ammonite Zones in North America. Part 1. Maryland and North Carolina: Bulletin of the American Museum of Natural History, no. 287.

Landman, N.H., Johnson, R.O., Edwards, L.E. (2004b). Cephalopods from the Cretaceous/Tertiary Boundary Interval on the Atlantic Coastal Plain, with a Description of the Highest Ammonite Zones in North America. Part 2. Northeastern Monmouth County, New Jersey: Bulletin of the American Museum of Natural History, no. 287, 1–107.

Landman, N.H., Johnson, R.O., Garb, M.P., Edwards, L.E., Kyte, F.T. (2007). Cephalopods from the Cretaceous/Tertiary Boundary Interval on the Atlantic Coastal Plain, with a Description of the Highest Ammonite Zones in North America. Part III. Manasquan River Basin, Monmouth County, New Jersey: Bulletin of the American Museum of Natural History, no. 303, 1–122.

Landman, N.H., Johnson, R.O., Garb, M.P., Edwards, L.E., Kyte, F.T. (2010). Ammonites from the Cretaceous/Tertiary Boundary, New Jersey, USA; in Tanabe K., Shigeta Y., Sasaki T., Hirano H. (eds.), *Cephalopods: Present and Past*. Tokai University Press, Tokyo, 287–295.

Lewis, J., Harris, A.S.D., Jones, K.J., Edmonds, R.L. (1999). Long Term Survival of Marine Planktonic Diatoms and Dinoflagellates in Stored Sediment Samples. *Journal of Plankton Research*, 21, 343–354.

Macellari, C. (1988). Stratigraphy, Sedimentology, and Paleoecology of Upper Cretaceous/Paleocene Shelf-deltaic Sediments of Seymour Island in Geological Society of America Memoir 196.

Macellari, C., Zinsmeister, W.J. (1983). Sedimentology and Macropaleontology of the Upper Cretaceous to Paleocene Sequence of Seymour Island. *Antarctic Journal of the U.S., Annual Review*, 18(5), 69–71.

Machalski, M. (2002). Danian Ammonites: A Discussion. *Bulletin of the Geological Society of Denmark*, 49, 49–52. https://doi.org/10.37570/bgsd-2003-49–05.

Machalski, M. (2005). Late Maastrichtian and Earliest Danian Scaphitid Ammonites from Central Europe: Taxonomy, Evolution, and Extinction. *Acta Palaeontologica Polonica*, 50(4), 653–696.

Machalski, M., Heinberg, C. (2005). Evidence for Ammonite Survival into the Danian (Paleogene) from the Cerithium Limestone at Stevns Klint, Denmark. *Bulletin of the Geological Society of Denmark*, 52, 97–111.

Machalski, M., Jagt, J.W.M., Landman, N.H., Motochyrova-Dekova, N. (2007). The Highest Records of North American Scaphitid Ammonites in the European Maastrichtian (Upper Cretaceous) and Their Stratigraphic Implications. *Acta Geologica Polonica*, 57(2).

MacLeod, N. (1998). Impacts and Marine Invertebrate Extinctions. *Special Publications of the Geological Society of London*, 140(1), 217–246.

MacLeod, N., Rawson, P.F., Forey, P.L., Banner, F.T., Boudagher-Fadel, M.K., Bown, P.R., Burnett, J.A., Chambers, P., Culver, S., Evans, S.E., Jeffery, C., Kaminski, M.A., Lord, A.R., Milner, A.C., Milner, A.R., Morris, N., Owen, E., Rosen, B.R., Smith, A.B., Taylor, P.D., Urquhart, E., Young, J.R. (1997). The Cretaceous–Tertiary Biotic Transition. *Journal of the Geological Society*, 154(2), 265–292.

Officer, C., Page, J.L. (1996). *The Great Dinosaur Extinction Controversy*. Helix Books, New York.

Petersen, S.V., Dutton, A., Lohmann, K.C. (2016). End-Cretaceous Extinction in Antarctica Linked to Both Deccan Volcanism and Meteorite Impact via Climate Change. *Nature Communications*, 7, 12079.

Powell, J.L. (1998). *Night Comes to the Cretaceous: Dinosaur Extinction and the Transformation of Modern Geology*. St. Martin's Press, New York.

Rasmussen, J.A., Heinberg, C., Hakansson, E. (2005). Planktonic Foraminifers, Biostratigraphy and the Diachronous Nature of the Lowermost Danian Cerithium Limeston at Stevns Klint, Denmark. *Bulletin of the Geological Society of Denmark*, 52, 111–131.

Ribeiro, S., Berge, T., Lundholm, N., Andersen, T., Abrantes, F., Ellegaard, M. (2011). Phytoplankton Growth after a Century of Dormancy Illuminates Past Resilience to Catastrophic Darkness. *Nature Communications*, 2, 311.

Robertson, D.S., McKenna, M.C., Toon, O.B., Hope, S., Lillegraven, J.A. (2004). Survival in the First Hours of the Cenozoic. *GSA Bulletin*, 116(5–6), 760–768.

Slattery, J.S., Harries, P.J., Sandness, A.L. (2012). A Review of Late Cretaceous (Campanian and Maastrichtian) Heteromorph Ammonite Paleobiology, Paleoecology, and Diversity in the Western Interior of North America; in Cavigelli, J.P. (ed.), *Invertebrates: Spineless Wonders, 18th Annual Tate Conference*. Tate Geological Museum, Casper College, Casper: WY, 76–93.

Vellekoop, J., Van Tilborgh, K.H., Van Knippenberg, P., Jagt, J.W.M., Stassen, P., Goolaerts, S., Speijer, R.P. (2019). Type-Maastrichtian Gastropod Faunas Show Rapid Ecosystem Recovery Following the Cretaceous-Palaeogene Boundary Catastrophe. *Palaeontology*.

Ward, P.D., Kennedy, W.J., MacLeod, K.G., Mount, J.F. (1991). Ammonite and Inoceramid Bivalve Extinction Patterns in Cretaceous/Tertiary Boundary Sections of the Biscay Region (Southwestern France, Northern Spain). *Geology*, 19(12), 1181–1184.

Witts, J.D., Landman, N.H., Garb, M.P., Irizarry, K.M., Larina, E., Thibault, N., Razmjooei, M.J., Yancey, T.E., Myers, C.E. (2021). Cephalopods from the Cretaceous-Paleogene (K-Pg) Boundary Interval on the Brazos River, Texas, and extinction of the Ammonites. *American Museum Novitates*, 2020(3964), 1–52.

Glossary

A

Acanthoceras: A genus of ammonites from the Cretaceous that are characterized by bold ribbing. Acanthocerids are commonly found relatively inshore compared to other ammonites.

agoniatitic: Of or relating to agoniatitids; it is most commonly used in reference to the sutures of agoniatitids, e.g., "agoniatitic suture."

agoniatitid: An ammonoid in the order Agoniatitida.

Agoniatitida: The most primitive order of ammonoids.

Agoniatites: The type genus of the order Agoniatitida.

Albian: The last stage of the Early Cretaceous, spanning about 113 to 100.5 Ma.

Albian-Cenomanian Boundary: The "minor extinction" which occurred 100.5 Ma, resulting in the end of the Albian and beginning of the Cenomanian stage.

allometry (allometric growth): A manner of ontogenetic development of organisms in which body proportions change throughout the growth cycle.

ammolite: Gem-quality iridescent ammonite fossils from the Bear Paw Formation of Canada.

ammonite: An ammonoid cephalopod belonging to the order Ammonitida. Ammonites have ammonitic sutures. While all ammonites are ammonoids, not all ammonoids are ammonites.

ammonite pavement: A quality of flat, weathered limestones in which ammonite fossils are so abundant that they appear to tile the outcropping. Examples include the "Ammonite Pavement" at Lyme Regis in Britain, Dalle Aux Ammonites in France, and Ammonitico Rosso in Italy.

ammonoid: An animal belonging to the phylum Mollusca, class Cephalopoda, and subclass Ammonoidea. Ammonoids lived from the Early Devonian Period to the earliest Paleocene Period.

anaptychus: The upper beak valve of an ammonoid (plural *anaptychi*). In life, an ammonoid had two anaptychi forming the upper beak and two aptychi forming the lower beak.

Anarcestes: A genus of early agoniatitids which appeared in the Early Devonian.

anarcestid: A member of the Anarcestidae, a family of very primitive agoniatitid ammonoids.

Ancyloceratina: The suborder of ammonites descended from *Tropaeum*, consisting of the true heteromorphs ammonites.

Ancyloceras: An early heteromorph appearing 129.5 Ma in the Late Jurassic and undergoing three distinct ontogenetic morphs: an open coil, a straight shaft, and a hook as maturity is reached. The type genus of suborder Ancyloceratina, ranging from about 10 cm to a meter in length.

ancylocone: A common heteromorphic shell shape which begins in an open coil (gyrocone), undergoes an intermediate phase of straight growth, and a final adult phase that is hooked.

Anomalocaris: An arthropod-like, free-swimming predator from the Early and Middle Cambrian.

Anthropocene: A proposed novel geologic period concurrent with the beginning of the Industrial Revolution, characterized by human-induced climate and ecological changes.

anthropogenic: Induced or caused by humans; a byproduct of human activity.

aperture: The opening in the shell through which the animals' soft tissue emerges.

apex: The shell's pointy starting point.

aptychus: The lower beak valve of an ammonoid (plural *aptychi*). In life, an ammonoid had two anaptychi forming the upper beak and two aptychi forming the lower beak.

aptychus-in-situ: An aptychus preserved in place; in the living chamber of the ammonoid shell.

Arthropod: Members of the phylum Arthropoda: invertebrates characterized by exoskeletons and jointed appendages. Arthropods have existed since at least the beginning of the Cambrian.

B

bactritid: A member of the order Bactritida.

Bactitida: An order of cephalopods that is transitional between nautiloids and their daughter taxa: the ammonoids and coleoids.

Bactrites: A common bactritid cephalopod.

Baculites: The type genus of straight-shelled heteromorph ammonites which became abundant in the Cretaceous Period.

baculitid: Members of the family Baculitidae: straight-shelled heteromorph ammonites from the Cretaceous Period.

belemnite: The fossilized internal shell of the belemnoid, a squid-like cephalopod from the Mesozoic.

biodiversity: The amount of variation of species in a given ecosystem, usually expressed as a number or genera or species.

body chamber: Also called the "living chamber." The last chamber of the ammonoid or nautiloid shell, in which the animal's soft tissue is housed (as opposed to the phragmocone chambers, which house cameral fluids and gases).

bottleneck event: Response by a given taxon to an extinction event in which precious few individuals survive, but they proliferate an explosive recovery of biodiversity.

brevidome: A body chamber length generally 180° (half a volution around the shell) or less.

buffalo stone: A term used by members of the Blackfoot Nation for ammolite, which is used as a talisman for successful hunts.

bullae: A type of micro-ornamentation evolved by Jurassic ammonites consisting of fine, delicate ribbing.

C

Cabrieroceras: One of the most common agoniatitid ammonoids from North America.

cadicone: Describes an ammonite shell whose aperture is wider than it is high, but not so extreme as a sphaerocone.

calcareous: Constructing a shell from calcium carbonate.

Callovian: A stage in the Middle Jurassic spanning from 166.1 to 163.5 Ma.

cameral fluid: Fluid held inside of the chambers of, or expelled from, cephalopod phragmacones.

carbon: The sixth periodic element, which plays several key roles in ecological systems.

carbonate: Sedimentary rocks composed of carbonate minerals, e.g., calcium carbonate.

carbonate platform: A sedimentary structure formed from continuous deposition of carbonates from calcareous organisms, such as a coral reef.

carbon dioxide: A gas whose molecules are composed of one carbon and two oxygen atoms. When it accumulates, carbon dioxide is an effective greenhouse gas.

Carboniferous: The geologic period stretching from 359.2 to 299 Ma, named for the coal swamps whose strata characterize this time interval.

Cenomanian: The first stage of the Late Cretaceous Period, lasting from about 100.5 to 93.9 Ma.

Cenomanian-Turonian Boundary: A "minor extinction" in the Late Cretaceous which involved rapid warming, mass dysoxia, and heightened salinity in the world's oceans. This was one of the most devastating events for ammonites during the Cretaceous Period.

Cenozoic: The third and current era of the Phanerozoic Eon, sometimes called the Age of Mammals.

center of gravity: The point in an organism's body at which it is stable. In ammonoids, the center of gravity is controlled largely by the length of the body chamber and determines the orientation of the shell.

cephalopod: A free-swimming, usually predatory mollusk of the class Cephlaopoda. Cephalopods include nautiloids, ammonoids, octopodes, and squid.

***Cephalopodenkalk*:** A lithofacies of certain fossiliferous limestones in which cephalopod shells are the dominant fossil.

ceratitid: A member of the ammonoid order Ceratitida which lived during the Triassic Period and are identifiable by the ceratized suture, in which saddles are smooth and parabolic, and lobes are serrated.

ceratized ammonite: A Jurassic ammonite whose sutures have converged on ceratitic sutures. They are normally easy to distinguish because the serrated lobe is parabolic and not straight.

chalk: A fine-grained limestone made of the tests of calcareous algae.

chamber: A hollow space inside the shell of a cephalopod.

clavi: Elliptically shaped, nodular ornamentations on the exterior of ammonite shells.

climate: The established meteorological regime for a prolonged period, often lasting seasons, years, or centuries.

climate change: Long-term cooling, warming, humidifying, or aridifying of the existing climate. It can be caused by biological, astronomical, and geological factors.

clymeniid: A member of the short-lived ammonoid order Clymeniida which lived in the Late Devonian and had a dorsal siphuncle.

Clymenia: The type genus of the Clymeniidae, which has a dorsal siphuncle, asymmetrical lateral lobes, and a highly evolute, nearly serpenticone coil.

coccolithophore: The microscopic marine algae whose calcareous shells form chalk deposits.

coleoid: A member of the cephalopod order Coleoidea, which includes octopodes, squid, cuttlefish, and belemnoids.

Coniacian: A stage of the Late Cretaceous Period which stretches from about 89.8 to 86.3 Ma.

costae: Also called "collars." Sharp ribs which grow intermittently through ontogeny of some ammonoid species, which may weather away, leaving lirae.

Cretaceous: The third and final geologic period in the Mesozoic Era.

Cretaceous-Tertiary Boundary: The mass extinction event which killed the non-avian dinosaurs 66 Ma.

D

Danian: The earliest stage of the Paleocene Epoch, which began the moment the Cretaceous Period ended 66 Ma and lasted until 61.6 Ma.

Demersal: Living of having a habitat close to the sea floor.

Devonian: The geologic period in the Late Paleozoic, which lasted from 419.2 to 358.9 Ma. It is preceded by the Silurian and succeeded by the Carboniferous.

derived: Describing a trait which evolved later than another, often for a more specialized function.

Didymoceras: A genus of demersal heteromorph ammonites from the Late Cretaceous with turricones and dramatic hooks.

distribution: The full geographic range of a given taxon.

diplomoceratid: A member of the Cretaceous heteromorph ammonite family Diplomoceratidae. Commonly called the "paperclip ammonites"; diplomoceratids can include other shapes and highly diverse suture geometry.

Discoscaphites: A genus of conservative heteromorph ammonites which are known to have survived the Cretaceous-Tertiary Boundary.

disparity: The degree of anatomical difference among species in a given ecosystem.

diversity: The amount of different species in a given ecosystem.

E

ectocochleate: Having an external shell.

embryo: The earliest phase of life for a cephalopod in which the umbilicus comprises the entire shell.

encrusting: (of a sedentary organism) to build a permanent exoskeleton on top of another surface; often another animal.

endocerid: A member of the orthocone nautiloid order Endocerida, which include the largest shelled cephalopods of all time.

endocochleate: Having an internalized shell.

epeiric sea: See epicontinental sea.

epicontinental sea: A body of salt water which has formed on a continent due to sea level rise.

epizoan: A sedentary animal which encrusts on top of another animal, such as a barnacle.

Eubostrychoceras: A genus of helically coiling heteromorph ammonites from the Cretaceous Period that are completely free of nodes or tubercles.

Eumorphoceras: A genus of goniatitids from the Carboniferous Period.

external lobe: Also called the "ventral lobe" or "(E)." The lobe which centers on the ventral edge, or keel, of the ammonoid shell.

extinction: The death of all members of a species. Often connotes a cause of this death.

F

facies: The immediately visible characteristics of a given rock outcrop. Can include biofacies, which describe the types of fossils, or lithofacies, which describe the rock itself.

Famennian: The final stage of the Devonian Period from 371.1 to 359.3 Ma. It is the second of two stages of the Late Devonian.

faunal assemblage: The fossil taxa present in a given outcrop or excavation.

faunal succession: The progression of species upward through the rock record, which may include the first-appearance datum, last-appearance datum, and increasing complexity or specialization.

Fibonacci series: A mathematical sequence of values which represent the "Golden Spiral," a pattern which is found in myriad natural structures.

filament: Thin, stringy appendages of some soft-bodied invertebrates, including cnidarians and cephalopods.

foram: See foraminifera.

Foraminifera: Amoeboid protists which secrete hard exoskeletons that are easily preserved in the microfossil record.

fractal: A mathematical phenomenon in which the part is identical to the whole. Fractal surfaces can be *true fractals*, which are not able to exist in three dimensions, or *natural fractals*, which simply mimic fractal geometry: irregular surfaces in the three-dimensional world mimic the hypothetical fractal model and can be measured the same way a fractal would be measured.

Frasnian: The first of two stages in the Late Devonian Period, and the penultimate stage of the Devonian Period.

Frasnian-Famennian Boundary: See Hangenberg Event.

G

Gastropod: Also called "univalve"; the class of mollusks that include snails and slugs.

generating curve: The imaginary flat surface that represents the angle, versus the apex, of new shell growth at a given point in an ammonite's ontogeny. It is parallel to the aperture.

Girtyoceras: A significant genus of goniatitids from the Carboniferous Period.

Givetian: A stage in the Middle Devonian lasting from 387.7 to 382.7 Ma.

glass sponge: Also called Hexactinellid sponges, which are hard sea sponges whose exoskeletons are constructed from siliceous pointed clasts.

goniatitid: Informally referred to as "goniatite," members of the ammonoid order Goniatitida. They are characterized by the goniatitic suture pattern.

Goniatites: The type genus of Goniatitida.

Great Dying: See the Permian-Triassic Boundary.

greenhouse gas: Gases which hold onto heat, and when sequestered in Earth's atmosphere, result in global warming. Common greenhouse gases include carbon dioxide, methane, and ozone.

greenhouse state: A prolonged period in which greenhouse gases have accumulated to the point that global temperatures are significantly increased.

gyrocone: Similar to a planispiral coil, except that the whorls are open and do not touch.

H

Hangenberg Event: The Devonian-Carboniferous Boundary; a mass extinction marked by sudden sea level fall and dysoxia which devastated ammonoids and other animals 359 Ma.

Helcionellid: Small shelled animals which emerged in the Early Cambrian and persisted into the Ordovician. Generally accepted as the first crown-group gastropods.

Helicoprion: A cartilaginous fish related to sharks from the Permian whose lower jaw spiraled in a saw-like whorl.

Harpoceras: A hildoceratid ammonite genus common in the Jurassic of Europe with a defined keel and bold, biconcave (zig-zagging) ribs.

heteromorph: Usually refers to members of the Suborder Ancyloceratina but is sometimes more broadly used to describe any ammonite whose shell coils in an aberrant or non-planispiral way.

hildoceratid: A member of the highly prolific Jurassic ammonite family Hildoceratidae.

Hildoceras: The type genus of the Hildoceratidae. They have concave, almost triangular ribs and are highly evolute despite being very laterally compressed. The genus and family are named after St. Hilda.

Hokkaido: The second-largest island in the Japanese archipelago which is known for its abundance and diversity of Cretaceous ammonites.

***Hoploscaphites*:** A genus of small, conservative heteromorphs (scaphitids) which have a low number of node rows and a somewhat open hook in maturity.

I

icehouse state: A period of global cooling brought on by a lack of atmospheric greenhouse gases.

ichthyosaur: A marine reptile similar in body shape to a dolphin that was common in the Jurassic. Ichthyosaurs gave live birth and fed on ammonoids and other cephalopods.

igneous province: A massive expanse of igneous extrusive rock, commonly associated with supervolcanism, tectonic activity, or mantle plumes. Large igneous provinces can accelerate global warming due to the release of greenhouse gases and have been a likely culprit in several significant extinction events.

index fossil: A fossil used by paleontologists and sedimentary geologists to delineate geological time. To be considered an index fossil, the fossil must be abundant, have geographically widespread distribution, quickly go extinct (the shorter the temporal range, the better), and be clearly visually different from other, similar species.

iniskim: See buffalo stone.

inoceramid: A type of large marine bivalve which cohabited epicontinental seaways with ammonites. Inoceramids are used to determine the baseline stable-isotope ratios for benthic and, by extension, pelagic species.

intraspecific variation: The degree of morphological disparity within a given species.

isometry (isometric growth): A manner of ontogenetic development of organisms in which body proportions remain consistent throughout the growth cycle.

isotope: An element may have two or more atomic weights assigned to a given chemical element based on the number of protons, which have no charge.

Izumi Group: Marine deposits on the Japanese island Honshu from the Upper Cretaceous, named for the Izumi Mountains.

J

***Jeletzkytes*:** See *Hoploscaphites.*

K

Kellwasser Event: The Frasnian-Fammenian extinction event that devastated numerous marine taxa, including ammonoids.

***Kepplerites*:** A small stephanoceratoid ammonite which defines the Callovian Stage of the Jurassic Period in Europe. Russian specimens are frequently pyritized.

***Kimberella*:** An animal from the Ediacaran Fauna that may represent the first mollusk.

Kimmeridgian: A stage from the Late Jurassic spanning 157.3–152.1 Ma.

***Kossmoceras*:** A Callovian ammonite exhibiting a high degree of ornamentation and sexual dimorphism.

L

lateral lobe: Abbreviated (L). The next lobe on either side of the External Lobe.

limestone: A biochemical sedimentary rock made from dissolved and precipitated calcium carbonate, generally from the dissolved shells of calcareous invertebrates.

living chamber: See body chamber.

living chamber length: The angle formed by the aperture and the back wall of the living chamber around the axis of coiling. Can be longidome, mesodome, or brevidome.

lobe: Septal folds which point toward the shell's apex.

longidome: A very long body chamber that extends more than 360° around the shell.

M

macroconch: Abbreviated [M]; the larger size class of conches in a given ammonoid species, which represents the presumed female.

***Manticoceras*:** One of the most significant genera of Late Devonian goni-
atitids. Manticoceras survived the Kellwasser Event and are highly
intraspecifically varied.

***Meekoceras*:** Ceratitids from the Triassic that are common at the famous
Union Wash locality.

mesodome: A moderate body chamber length that usually extends 260°–
300° around the shell.

Mesozoic: The second era of the Phanerozoic Eon, sometimes referred to
as the Age of Reptiles. Mesozoic means "middle life" and extends
temporally from 252 to 66 Ma. The Mesozoic Era is composed of the
Triassic, Jurassic, and Cretaceous Periods.

microconch: Abbreviated [m]; the smaller size class of conches in a given
ammonoid species, which represents the presumed male.

micro-ornamentation: Very fine relief patterns on Mesozoic ammonites
from the Jurassic and Cretaceous that commonly fall into bullae,
nodes, tubercles, and clavi.

minor extinction: Extinction events of moderate rank, which are less
severe than true mass extinction events, but more severe than the
continuous, gradual process of "background extinction."

mollusk: A member of the phylum Mollusca, which includes
gastropods (snails and slugs), bivalves (clams, mussels, and scallops), monoplacophorans, chitons, and cephalopods. After the
arthropods, Mollusca represents the next largest phylum for
invertebrates.

morph: An age-segregated growth phase in the life cycle of an ammonoid.
Coiling and ornamentation can change drastically and often indicate
different behavioral niches over ontogeny.

N

nautiloid: Cephaolopods in the subclass Nautiloidea. Nautiloids are the
ancestor of the bactritids, ammonoids, and coleoids.

nautilus: Nautiloid cephlaopods that are members of the family Nautilitidae.

***Nautilus*:** Nautiloid cephlaopods that are members of the genus *Nautilus*.

nectonic: Free-swimming.

nectobenthic: Moving freely on or above the sea floor.

***Nektocaris*:** A Burgess shale invertebrate with traits loosely similar to both
mollusks and arthropods, but its siphon was once got it considered

a possible stem-group cephalopod. It is no longer believed to be a cephalopod ancestor.

neritic: A moderately inshore zone of the ocean that exists primarily on the continental slope and has a depth up to 200 m.

node: A small, round type of ornamentation on the exterior of ammonoid shells, commonly occurring in bilaterally symmetrical rows. Nodes also cap off longer tubercles and spines so that if they break, the animal's shell does not spring a leak.

nostoceratid: A member of the heteromorph ammonite family Nostoceratidae. Nostoceratid shells are extremely diverse and account for most of the shapes heteromorph ammonites evolved, as well as most of the bizarrest ones.

***Nostoceras*:** The type genus of the Nostoceratidae which lived in the Late Cretaceous and is the probable parent taxon to *Didymoceras, Pravitoceras, Nipponites,* and other distinctive heteromorph ammonites. *Nostoceras* had abundance on several continents and its name, which means "returning horn," comes from the hook it forms in the final growth phase.

O

ontogeny: The progression of anatomical changes to an animal over its lifecycle.

***Ophiceras*:** The Triassic ammonoid from which all Jurassic and Cretaceous ammonoids are derived.

ornamentation: Patterns of texture on a mollusk shell's exterior.

***Otoceras*:** One of three ammonoid genera that survived the Permian-Triassic extinction.

oxycone: A fairly laterally compressed planispiral.

***Oxybeloceras*:** A genus of small diplomoceratid ammonites.

ozone layer: A layer of Earth's stratosphere dominated by the gas ozone, which acts as a shield against most harmful solar radiation.

P

paleoecology: The environmental interactions of organisms from the fossil record.

paleoenvironment: The biophysical, climatic, and geological environment which existed at a given time in the prehistoric past. Paleoenvironments are usually described through the faunal assemblages and expected paleoecology of the fossil organisms present.

Paleozoic: The earliest era of the Phanerozoic Eon which spans the time between the onset of the Precambrian and the end of the Permian, about 538.8–252 Ma.

paralarval: A stage in cephalopod development or ontogeny between hatchlings and subadults. Paralarvae are not true larvae because they do not undergo metamorphosis, merely allometric growth.

Perisphinctes: A genus of perisphinctid ammonites from the Middle and Late Jurassic period characterized by medially invaginated bullae and moderate evoluteness. *Perisphinctes* is a common index fossil for parts of the Late Jurassic.

perisphinctid: A member of the Jurassic ammonite family Perisphinctidae. Perisphinctids are derived from stephanoceratids.

Perisphinctoidea: The superfamily of ammonites that includes perisphinctids, dorsoplanitids, and other common ammonite families spanning the Middle Jurassic to Early Cretaceous.

Permian: The final geologic period of the Paleozoic Era, which lasted from 289.9 to 252 Ma.

Permian-Triassic Boundary: Changes in the rock and fossil records at the end of the Permian Period which suggest the greatest mass extinction of all time.

phragmocone: The portion of a cephalopod's shell that is comprised of buoyancy chambers, not the living chamber.

Phylloceras: A somewhat basal genus of ammonites characterized by involute coiling and oak-leaf saddles and lobes, which appeared in the Early Jurassic and persisted into the last stage of the Cretaceous.

phylloceratid: Members of the ammonite family Phylloceratidae.

phylogeny: The branching sequence of relationships between increasingly derived taxa.

phylogenetic tree: A "family tree" created for genera and species based on faunal succession and the identification of primitive and derived traits.

phylum: The taxonomic classification that ranks below kingdom and above class. Cephalopods are in the phylum Mollusca.

Pierre Shale: A unit of marine sedimentary rock in the middle of North America that is dated to the Upper Cretaceous. The Pierre Shale contains numerous ammonite species as well as fish, marine reptiles, inoceramids, and countless other taxa.

Placenticeras: A genus of large oxycone ammonites from the Upper Cretaceous with many small umbilical lobes. *Placenticeras* are common in the Pierre Shale and Bear Paw Formations, and they are one of the most common producers of ammolite.

planispiral: A normally coiling ammonite or nautiloid that has a closed coil (some amount of contact between whorls).

planktic: An organism whose life habit comprises drifting in open water.

platycone: The most laterally compressed planispirals.

Plectronoceras: A genus of mollusks commonly considered the first true cephalopod.

primitive: Having traits which evolved earlier in a lineage or succession; usually which are not particularly specialized.

Prolecanitida: An order of ammonoids that is considered transitional between goniatitids and ceratitids. Prolecanitids have backward-pointing septal necks and sutures that are normally intermediary between goniatitic and ceratitic.

pseudoplanktonism: The life habit of an organism which attaches to a floating object semi-permanently or permanently.

R

radiation: (evolution) The redistribution of new species, especially after a major environmental change or a pulse of extinction and recovery, e.g., "adaptive radiation."

reef: A sedentary structure created by encrusting organisms underwater which affect water flow at the surface without touching it. Reefs provide homes and food for other species.

regression: An event in which sea level falls, and the shoreline migrates outward. Regressions can either be local or global, and common causes include global cooling or tectonics.

S

saddle: A fold within an ammonoid suture that points toward the aperture.

salinity: The concentration of salt in seawater. Salinity is greater at the equator than the poles because warm water has a greater capacity for holding NaCl molecules.

Santonian: Stage of the Late Cretaceous, spanning the interval from 83.6–86.3 Ma.

sedentary: Describes an animal that does not move. Sedentary animals usually encrust their exoskeletons. Barnacles, corals, and rudists are all examples of sedentary animals.

sedimentary: The group of rocks that are formed by pieces of other rocks, usually through the processes of weathering and erosion.

sedimentology: The study of sedimentary rocks and the process of sedimentation.

septa: The thin walls of calcium carbonate separating the chambers of an ammonoid or nautiloid shell.

septal neck: The straw-like protrusions that come out of the septa to lend structural support to the siphuncle.

septal wall: See septa.

shale: A fine-grained siliciclastic sedimentary rock representing a moderate or deep marine environment.

silicate: Rocks and minerals that are abundant in silicon dioxide.

siphuncle: The fleshy tube that is threaded back through the phragmo-cone of ammonoids, emptying or filling the chambers of cameral fluid.

***Solenoceras*:** A genus of small diplomoceratid ammonites from the Cretaceous Period.

spine: A hollow cone of shell that is capped off inside by a hollow node in case the elongated spine should break off.

strata: Layers of rock in nature, particularly as they are deposited horizontally through igneous and sedimentary processes.

stratigraphy: The geological discipline which focuses on the order in which rock layers were deposited as well as their position, especially as it relates to geological history.

suture: The pattern formed at the intersection of a cephalopod shell's outer wall and a septum.

sutural complexity: The overall shape and amount of folding in an ammonoid's suture pattern.

T

Tithonian: The last stage of the Jurassic Period, which lasted from 152.1 to 144 Ma.

Toarcian: A stage in the Early Jurassic Period from 182.7 to 174.1 Ma.

***Tornoceras*:** A goniatitid from the Middle and Late Devonian which sometimes had unusually complex sutures.

transgression: An event in which sea level rises, and the shoreline migrates inland. Transgressions can either be local or global, and common causes include global warming or tectonics.

Triassic: The first geologic period of the Mesozoic Era, which lasted from 252 to 201 Ma.

Triassic-Jurassic Boundary: The end-Triassic extinction event about 201.3 Ma which was caused by intense volcanism.

tubercle: A short spine on an ammonoid shell.

turricone: A tight helically coiling ammonoid shell shaped like a long cone.

***Turrilites*:** A genus of helically coiled heteromorph ammonites from the Late Cretaceous.

turrilitid: A member of the family Turrilitidae, heteromorph ammonites.

U

umbilical lobe: Any lobe in an ammonoid suture which is not the external or lateral lobe. Umbilical lobes are farther from the center than both the external and lateral lobes and can range in number from one family of ammonoids to another.

umbilicus: The starting point of the shell, located at the shell's apex. For planispirals, this is in the center of the coil. The umbilicus represents the embryonic growth stage of the animal.

V

Vienna Peedee Belemnite: Also called the VPDB. A fossil belemnite from the Cretaceous Peedee Formation whose carbon isotopes indicate it was extremely pelagic, and therefore lived high in the water column. Though the fossil itself no longer exists, the stable-isotope ratios of the VPDB are used as a baseline for pelagic mollusks, as opposed to the stable-isotope ratios of benthic mollusks from the same time, which are compared to inoceramids.

vermicone: An ammonite shell that is knotted or "worm shaped" in appearance.

vermetid: A group of marine snails whose shells are extremely aberrantly coiled. Vermetid snails have been compared to heteromorph ammonites, but they do not float or have chambered shells and the comparison is considered outdated.

volcanism: The process by which molten material (igneous extrusive) is ejected onto the Earth's surface.

volution: See whorl.

W

Watinoceras: A Late Cretaceous ammonite that helped delineate the onset of the Turonian Stage.

Western Interior Seaway: An epicontinental sea which formed in the middle of North America and lasted through most of the Cretaceous Period. It is a common source of marine fossils from North America, including ammonites.

whorl: (Of a coiled shell) A full 360-degree rotation/growth cycle around the axis of coiling.

X

Xenodiscus: One of three genera of ammonoids to survive the Great Dying. All Jurassic and Cretaceous ammonite lineages can be traced to its descendant *Ophiceras*.

xenomorph: An impression of a foreign object in the shells of encrusting bivalves.

Y

Yezo Supergroup: An Upper Cretaceous sedimentary unit in Japan from which numerous heteromorph ammonite species are known.

Index

Milton Keynes UK
Ingram Content Group UK Ltd.
UKHW052027141024
449569UK00016B/728